The Birth of
Top 40 Radio

The Birth of Top 40 Radio

The Storz Stations' Revolution of the 1950s and 1960s

RICHARD W. FATHERLEY *and*
DAVID T. MACFARLAND

Forewords by
Deane Johnson *and* Bud Connell

McFarland & Company, Inc., Publishers
Jefferson, North Carolina, and London

All photographs courtesy Bud Connell.

LIBRARY OF CONGRESS CATALOGUING-IN-PUBLICATION DATA

Fatherley, Richard W.
    The birth of top 40 radio : the Storz stations' revolution of the 1950s and 1960s / Richard W. Fatherley and David T. MacFarland ; forewords by Deane Johnson and Bud Connell.
       p.   cm.
    Includes bibliographical references and index.

    ISBN 978-0-7864-7630-5
    softcover : acid free paper ∞

    1. Popular music radio stations—United States.
I. MacFarland, David T. II. Title.
PN1991.67.P67F38 2014
338.4'762—dc23   2013041031

BRITISH LIBRARY CATALOGUING DATA ARE AVAILABLE

© 2014 Richard Ward Fatherley, Jr., and David T. MacFarland. All rights reserved

No part of this book may be reproduced or transmitted in any form or by any means, electronic or mechanical, including photocopying or recording, or by any information storage and retrieval system, without permission in writing from the publisher.

On the cover: Sandy Jackson at KOWH (courtesy Bud Connell); radio and background images (iStockPhoto/Thinkstock)

Manufactured in the United States of America

McFarland & Company, Inc., Publishers
  Box 611, Jefferson, North Carolina 28640
    www.mcfarlandpub.com

Dedication

Dick Fatherley in 1968.

Richard W. Fatherley was the original author of this book, but he died of a heart attack before it could be completed.

It seems that few persons anymore are so enraptured, so motivated, or so inspired by the company they work for that they want to write a glowing book about it. Loyalty now, both down from the top and up from the ranks of most organizations, is in short supply. Buy-outs and mergers have made it impossible for employees to get very secure with their current position. It used to be that entire families — generation after generation — worked for the same company. "My dad, and his father before him, worked at GM, and now I do too" — be it in the mine, on the farm, or wherever. Workers had the feeling that the company would last forever, and would take good care of them after they had given it their youth, years of loyalty, and their trust.

The demise of radio was predicted in the middle of the last century by vaunted industrialists like General David Sarnoff of RCA, who stated flatly "Let's face it boys, radio is dead." During those dark days of radio in the early 50s, we youngsters were told "Anyone with any brains is going into TV." Radio networks collapsed. A lot of talent sought the glamor of television. But if radio was dead, they forgot to bury the body. By 1953 it was discovered that radio sets were the largest-selling electrical appliance in the U.S., and by a wide margin. At the end of that same decade, the number of radios being sold was increasing, not dwindling.

Into radio's early-50s malaise had come Todd Storz of Omaha. He would show the industry just how wrong Sarnoff was.

The Storz Broadcasting Company did not become successful because of their business practices, although they were very good. Instead, the Storz formula for success was a precursor of the saying "Build it and they will come." That is, provide appealing programming that generates high audience ratings, and the business will almost take care of itself.

Richard Ward Fatherley, who had worked as an announcer and as production director at Storz stations WHB and KXOK, was searching the country for facts about American radio during the last half of the 20th century. He had made it his job to tell the Storz Top 40 story as accurately and as interestingly as he could. Dick had completed 14 short chapters of his intended book before he died suddenly in 2010. His emphasis was on the programming heard on the seven Storz stations in Omaha, New Orleans, Kansas City, Minneapolis, Miami, Oklahoma City, and St. Louis, on how they developed the Storz "sound," and how they took the entire radio industry to a whole new level of listener loyalty.

Dick Fatherley's unfinished manuscript is remarkable because he felt compelled to express his love for a company that meant so much to him and his career. He wanted to explain the history-making impact, the innovation, and the lasting influence of the programming of the Storz Broadcasting Company–what Fatherley called "Radio's Revolution" — and how it helped to save the U.S. radio industry from oblivion in the early 1950s.

Following Dick's passing, four of us thought the story was too good not to be available to a wider audience. Our group was David MacFarland, whose

Ph.D dissertation was *The Development of the Top 40 Radio Format*; Bud Connell, one of the nation's most accomplished radio programmers and consultants; Deane Johnson, a former program director of several of the Storz stations; and Jack Sampson, a general manager of several Storz stations, and a more-than-20-year veteran of the Storz organization. We all pledged to continue the research and extend the writing in order to complete the project Dick had started.

It is not known to us exactly what the concluding chapters of Dick's book might have said. What in fact happened to the Storz Broadcasting Company is that after Todd Storz died in 1964, the company took a definite turn from an emphasis on programming, to being a tough, commercial, money-making machine. Partially because of that change, and the fact that all of the Storz properties were AM stations which were destined to lose their audience to the higher fidelity of FM signals, Storz Broadcasting began a steady decline until all of the stations were sold off.

The programming-driven rise and the avoidable fall of the Storz Broadcasting Company is not as important as the fact that the programming "sound" developed by Storz lives on today. When you tune to some AM stations, many FM signals, certain satellite radio channels, even Pandora on the internet, you hear Storz Broadcasting's influence. The real story you'll find in these pages is about innovations in radio programming that created — and pleased — entirely new audiences, and which ultimately spawned many successful variations on the original Storz format.

How many times in commencement speeches have speakers told grads to "follow your dreams"? Thank heaven Todd Storz did, and thank heaven Dick Fatherley did too.

— Jack Sampson

Acknowledgments

Grateful appreciation is extended for the help provided by:

Bud Connell, Deane Johnson and Jack Sampson for their innumerable contributions to this book. They offered unique and valuable printed and recorded material from their own archives, personal recollections not documented anywhere else, and unfailing collegiality throughout the writing and editing process. Their accomplishments in radio extend far beyond what this single book can convey.

Prof. Christopher H. Sterling, Research Professor of Media and Public Affairs in the Columbian College of Arts and Sciences at George Washington University, who patiently supplied both additional historical context and essential editorial advice. His counsel has been invaluable.

Kristin Copeland, former Senior Administrative Assistant in the A. Q. Miller School of Journalism and Mass Communications at Kansas State University, who graciously typed several revisions of Richard W. Fatherley's unfinished manuscript, thus making it available in electronic format.

Table of Contents

Dedication	v
Acknowledgments	viii
Foreword by Deane Johnson	1
Foreword by Bud Connell	3
Preface by David T. MacFarland	5
ONE • From Stowaways to Society, from Beer to Broadcasting	7
TWO • The Incubator: KOWH, Omaha	19
THREE • Forty Favorites in the Big Easy: WTIX, New Orleans	37
FOUR • Building the Flagship: WHB, Kansas City	55
FIVE • Signals from the Frozen North, the F.C.C. and the Sunny South: WDGY and WQAM	70
SIX • Programming Conventions I: Learning the Basics	89
SEVEN • Programming Conventions II: Tarnishing the Top 40, and Touting "Talk"	106
EIGHT • The Air War in Oklahoma City: KOMA VS. WKY	114
NINE • The Last Hurrah: KXOK, St. Louis	124
TEN • Elements of the Storz Station "Sound"	136
ELEVEN • Four Sages at Four Stages	154
TWELVE • The Decline, Sale and Legacy of Storz Broadcasting	178
Appendix: A Storz Broadcasting Timeline by Bud Connell	193
Chapter Notes	198
A Bibliographic Note	200
Index	201

Foreword

BY DEANE JOHNSON

I first became fascinated with radio before I was even in the seventh grade in the small agriculture community of Creston, Iowa. Our local station was KSIB, and it seemed like a magic place that was somehow bigger than life. While I was in junior high, KSIB's "man on the street" program, broadcast from a nearby business, occupied my attention during the noon-hour break from classes.

As I moved on to my first year in high school, a fellow named Todd Storz was experimenting with a daytime-only station in Omaha, Nebraska, well within reception range at Creston. They were playing some neat music and I became aware of a disc jockey named Sandy Jackson who was really cool. I was becoming hooked, big time. When the young people in Creston gathered around the filling stations and local eateries, it was KOWH that played on all of the car radios. This was the stuff dreams were made of. So I began hanging out at KSIB, making friends with the key personnel, even filling in once in awhile, and at the same time helping with a regular broadcast fed to the station from our high school. I think I was probably a royal pain in the neck to everyone.

Upon graduation from high school, I was so totally hooked that I knew where I was headed. It was radio. There was no doubt in my mind. I didn't care about anything else. I decided to go to engineering school in Kansas City, believing that a FCC First Class License was essential under the FCC rules of the day. I had no idea then how much that would play into my future goals.

The year of my move was 1954, and a visit to WHB proved to be spellbinding. I remember sitting in the studio while Wayne Stitt did his show. That was shortly after Todd Storz had purchased the station, and it had not then matured to its full glory. I did not yet know who Todd Storz was. But slowly, as Storz Broadcasting developed, and I listened to the new WHB from wherever I was working in Iowa, Missouri, Nebraska, or Kansas, I developed an overwhelming feeling that being part of Storz Broadcasting someday would be the ultimate accomplishment in radio — though probably not achievable. Little did I imagine that in 1961, after working in multiple small market stations, Storz National Program Director Grahame Richards would recommend to manager Jack Sampson that he hire me to program Storz's 50,000-watt KOMA in Oklahoma City. It was like going to heaven but remaining on earth.

Not only did I have the enviable opportunity to serve as program director of KOMA, but that was followed by being selected by Storz's executive vice-president Bud Armstrong to be program director of the Storz station where Top 40 had originally developed — WTIX in New Orleans. Lightning had struck twice.

The spirit of the Storz empire began to wither following Todd's untimely death, so I moved on to other pursuits in broadcasting. Those included programming NBC's 50,000-watt WKYC in Cleveland as they tried their hand at operating a Top 40 station, becoming program director for Don Burden's flagship KOIL in Omaha, then on to Minneapolis to program and manage KDWB against Storz's WDGY. After that, I moved out of programming and air work, and on to ownership and management of several stations, including a unique and dominant "full-service" Top 40 station in St. Joseph, Missouri, KKJO. Later, I would serve as president of two different multi-station national groups, but none of those responsibilities would ever approach the thrill of realizing the impossible dream of being part of Storz Broadcasting during its greatest years.

Deane Johnson programmed and managed radio stations for more than 40 years, beginning in 1961 at Todd Storz's KOMA in Oklahoma City. He also programmed Storz's WTIX in New Orleans. He has worked at NBC, KDWB (Minneapolis–St. Paul) and KOIL (Omaha), and was president of two station groups.

Foreword
BY BUD CONNELL

My friend of fifty years, Richard Fatherley began this project in the mid-1960s by interviewing broadcasting's current and former Storz employees, many of whom had become the industry's elite. A few years ago, he informed me he was exhuming his ancient files and would be writing what he hoped would be the definitive history of the Storz Broadcasting Company, the groundbreaking chain of radio stations developed by Todd Storz, the young man who created Top 40 and saved radio from death by TV.

Dick was about halfway along with his project in early 2010 when a heart attack took his life. He had been in weekly and sometimes daily contact with me, worrying over a subject or a turn of phrase, and I told his family I would make certain his book was finished. My first act was to recruit the talented "Doctor of Top 40" from Kansas State University, David MacFarland. He graciously accepted and in turn recruited the assistance of Deane Johnson and Jack Sampson, both longtime former Storz employees.

Dick Fatherley was a giant of a man, and one of the best broadcasters I've ever known. He could create a symphony of sound with nothing more than his mind, his big booming voice, and his talented fingers on the controls. Throughout his life, he was responsible for causing literally millions of dollars to change hands. Through the hundreds, perhaps thousands, of big name commercials he voiced for major corporations, he motivated commerce in this country as few people can. We need more like him, especially now.

I'll always miss my friend Dick, who'd announce himself on the phone with "Hi, Buddy, from the barbeque capital, Kansas City!"

Bud Connell was an on-air radio personality and also worked in programming and management consulting. Later he produced television programs, commercials, film and video for companies worldwide. He is now a novelist, penning award-winning thrillers.

Preface

BY DAVID T. MACFARLAND

This book is about a group of American radio stations that, beginning in the early 1950s, developed the Top 40 format into the dominant radio programming mode of the mid-twentieth century. Although other radio station group owners also achieved success by playing the most popular records of the day, the stations owned by the Mid-Continent Broadcasting Company are generally acknowledged as the pioneers of the Top 40 format. The company — known in its later years as The Storz Stations — was an innovator of radio programming practices and radio marketing techniques, and was hugely successful in attracting and holding audiences. Storz Top 40 programming achieved year-upon-year audience increases in the 1950s and early 1960s, even as radio network drama and comedy programs migrated to television — which in turn caused television sets to displace radios from the American living room. The catchphrase "Top 40 snatched radio from the jaws of television" may be overly-dramatic, but in the late 1950s and early 1960s, it was the truth.

Robert H. Storz and his son Todd were the owners of seven Top 40 stations that bore the family name. The first station — KOWH — was acquired in 1949, followed in succeeding years by facilities in New Orleans, Minneapolis, Kansas City, Miami, Oklahoma City, and St. Louis. KOWH was based in the Storz home town of Omaha, Nebraska, where the family also owned a successful brewery. After a few years of attempting to sound like the other stations in the market, young Todd began to pay attention to audience surveys that showed an increase in listenership whenever currently-popular records were aired. Storz expanded the number of hours when hit records were played, and ratings continued to climb, so that eventually hit recordings were being programmed all day long. An important operational principle had been established: Try something new, then check the ratings to see if audiences increased or dropped off. Growing audience numbers meant advertising rates could be raised, so more revenue flowed into the station. Top 40 radio was referred to in the trade as *format* radio, in which many disparate audio elements were aired each hour, based on a recurring format. In the popular press, it was often called "formula" radio, as if it were "brewed up" from the exact same recipe every day–but that was not the case at the Storz Stations in their prime. The essential element was always to "try something new"—a new record, a new set of station jingles, a new contest, a new promotion, a new disc jockey, etc. In their heyday, there was simply more "freshness" to the sound of Storz Top 40 stations compared to any other signals on the dial.

This book lays claim to being the definitive history of the Storz Stations in the Top 40 era because it is deeply informed by the people responsible for the advent and evolution of Storz Top 40 programming. The late Dick Fatherley was a production director and air personality at KXOK in St. Louis, and later was program director of WHB in Kansas City. When I took over the project after his death, I had invaluable contributions from three other knowledgeable insiders. Jack Sampson was sales manager at WHB in Kansas City, then managed KOMA in Oklahoma City, and finally became manager of KXOK in St. Louis. Deane Johnson was a program director at KOMA, and also at WTIX, New Orleans. Bud Connell was an air personality at KOWH, Omaha, and program director at KXOK — as well as competing against Storz at WFUN, Miami, and WNOE, New Orleans.

I never worked at a Storz station, but the joy of listening to Storz's WQAM in Miami during my late teen years, and working the afternoon drive slot one summer on a little daytime-only top 40 station in Cocoa, Florida, eventually led to my 642-page Ph.D. dissertation, later published by Arno Press as *The Development of the Top 40 Radio Format*.

Anyone who took part in this radio revolution would gladly revisit those days. Our book invites you to take the same journey.

• ONE •

From Stowaways to Society, from Beer to Broadcasting

The 1963 widescreen movie *It's a Mad, Mad, Mad, Mad World* is a raucous comedy that shows what happens when ordinary people get a chance to find hidden treasure—$350,000 in cash in 1960s dollars (about $2,592,459 in 2012)—buried under a mysterious "W" on a beach in Santa Rosita, California. The laughs are generated from the way wildly different characters hatch their treasure hunting plans and react to rivals, reversals, and both physical and social obstacles. Near the end of the movie, a suitcase containing all of the loot is brought to the top of a fire escape, where it accidentally pops open and scatters the bills to the spectators below.

By the early 1960s, the lure of free money—whether fluttering from the top of a downtown building or from the branches of a tree in a park—was "old news" to listeners in cities where a Storz radio station was operating. Those audiences—first in Omaha, then New Orleans, Kansas City, Minneapolis, Miami, Oklahoma City, and finally St. Louis—initially may have tuned in for the chance to grab some extra cash. But they continued to listen because from the mid–1950s into the 1970s, the Storz Broadcasting Company's radio stations were the most dynamic, exciting, fun-to-listen-to signals in America.

The essence of the station group's appeal wasn't easy money. (In fact, Todd Storz got ownership of a station in Miami only after promising the FCC *not* to run any treasure hunts.) Instead, the appeal was music—specifically the Top 40 format—which had been introduced at Storz's New Orleans station WTIX, and was then continuously fine-tuned. Other radio broadcasters noted that Storz stations were also extraordinarily profitable during a period when television was grabbing an ever-larger share of Americans' leisure time, and a correspondingly larger share of advertising dollars. Thus, it is not surprising that Storz programming practices, promotional strategies, and operational procedures were widely imitated by other radio stations across America.

The experiences of two illegal aliens who escaped from Europe without passports to seek their fame and fortune in America provide widely divergent context for the early history of the Storz radio broadcasting enterprise. The later immigrant would become the notorious manager of a world renowned singer, who used Top 40 radio to promote his records. The earlier expatriate would become a noteworthy midwestern industrialist and the grandfather of the creator of the Top 40 radio programming format.

Andreas Cornelis van Kuijk successfully marketed the wiggling frame and trembling singing voice of Elvis Presley, and propelled Presley's records to the top of the music charts and top-of-mind consciousness of 1950s and '60s young Americans.

In his *Elvis: A Radio History from 1945 to 1955*, Presley biographer Aaron Webster relates that "Andreas Cornelis van Kuijk [was] born in 1909 in Breda, Holland, an English-speaking country. At the age of twenty, he came to America — the land of opportunity — in a criminal way as a stowaway in a North Atlantic merchant ship. For free food and lodging, he immediately enlisted in the U.S. military and changed his name to Thomas Alan Parker ... 'Tom Parker' always claimed he was born in Huntington, West Virginia, and somehow, year after year, he got away ... with his disguise."

"In 1948," wrote Webster, "Parker was commissioned an honorary Colonel by Louisiana governor Jimmy Davis. He wanted to be addressed as 'Colonel' from then on. He sported the title like a *bona fide* southern gentleman. Those who did business with him say he was anything but." In 1945, he became country music star Eddie Arnold's manager. In 1953, Arnold fired Parker because he was also working with rival singer Hank Snow. In 1955, Parker became aware of an electrifying young singer named Elvis Presley. By 1956, Parker had moved Elvis from tiny Sun Records to giant RCA Victor, and had convinced Elvis to make him his exclusive agent. The rest of the Presley-Parker story is the Hollywood stuff from which stars and legends are created.

The earlier stowaway was Gottlieb Storz, of Benningen, a town along the Neckar River which runs through the Black Forest in southwest Germany and empties into the Rhine. Born January 21, 1852, Gottlieb was the seventh of ten children. Poverty forced the children to work outside the home — in addition to their daily chores — to bring in much needed extra money, especially after their father Johann died in 1862, at 46 years of age.

After completing his schooling, Gottlieb learned the trade of barrel-making and the brewing business. In 1872, twenty-year-old Gottlieb left for America, sidestepping his military obligation to the Kaiser. He found work in New York City as a cooper (barrel-maker) at a German brewery. He soon began a westward trek, taking the position of brewmaster for different companies in Philadelphia, Cincinnati, and St. Louis.

In 1876, Gottlieb Storz was hired as brewmaster by Joseph S. Bauman, owner of the Columbia Brewery in Omaha, Nebraska. Based on his success, Gottlieb later bought out the owner and went into business for himself as head of the new Storz Brewing Company. Gottlieb made his brewery operation one of the biggest in the United States prior to World War I. At 32 years of age, Gottlieb's Storz Beer brand was fast becoming the toast of the town.

Gottlieb Storz, Todd's grandfather. Born in southwest Germany in 1852, the seventh of ten children. At 20, he emigrated to America, working as a barrel-maker and later a brewmaster in Philadelphia, Cincinnati, St. Louis, and finally Omaha, Nebraska, where he went into business for himself as head of the new Storz Brewing Company.

The Storz Brewery survived the 1919–1932 Prohibition era by making "near-beer," ice cream, soda pop, and ice. Gottlieb Storz married wealthy widow Minna Buck from Pfullengen in 1882. Their union produced six children, one of whom was Robert Herman Storz, whose own son, Robert Todd Storz, is one of the principal subjects of this volume.

The six Storz brothers and sisters grew up in a stately home, completed in 1907, at 3708 Farnam Street along Omaha's so-called "Gold Coast," the colloquial address for Omaha's rich and famous. (To see the home online, search for "Gottlieb Storz House" on Wikipedia.)

A Personal Message of Welcome to Our Guests, a guide written by Arthur C. Storz, Jr., describes the opulence of the family mansion: "the solarium, modeled after the main dining room of the North German Lines luxury liner, the *Bremen*; the dazzling brass and copper chandeliers in the grand foyer entrance; the stunning crystal chandelier and sconces in the music room; the gorgeous Meissen, Dresden and Hummel china throughout the home; the grandfather clock with Westminster chimes that strike every fifteen minutes ... and the ballroom where Adele and Fred Astaire started their illustrious career."

Arthur C. Storz concluded his guide to the Storz mansion by declaring that it "speaks to a time when industrialists across America established labor intensive businesses through very great risks, hard work, and visionary spirit."[i]

In 1920, KDKA in Pittsburgh, Pennsylvania, announced Warren G. Harding had swamped James M. Cox with 60 percent of the popular vote to win the American presidency. For the first time, a handful of people could hear those results immediately by radio instead of waiting to read about the election in the morning newspapers.

Radio had arrived.

A year later in 1921, Robert H. Storz married Omaha native Mildred Todd. Robert H. Storz had served in World War I, then returned to Omaha to help his eldest brother Adolph run the Storz Brewing company. He also entered the banking business, "starting as a messenger, winding up as a Director," according to the West Omaha and Dundee Sun. Robert H. Storz and Mildred had two children, Robert Todd Storz, born May 8, 1924, and a younger sister, Susan. Reared in a fashionable three-story brick home near the University of Omaha (now the University of Nebraska at Omaha), Todd and Susan attended public schools.

"Mother was very pretty, always properly dressed, and perfect at entertaining people, which was the way Dad wanted it," said Susan. Mild-mannered Mildred was aloof and soft-spoken, with Todd inheriting much of her disposition. Susan's behavior was like her father's. At the dinner table, recalled Susan, "There was no discussion about what we did that day. It was all about what Dad did at business."

Life for Robert H. Storz centered on checks and balances, with no wish to experience even a hint of the storied poverty experienced by his orphaned father. A rigid pragmatist, Robert cultivated his growing interest in local associations and activities to position himself as a "mover and shaker." Together with brother Arthur, Robert H. Storz became instrumental in locating the Strategic Air Command (SAC) (under cigar-chomping four-star General Curtis E. LeMay) at Offutt Field in nearby Bellevue, Nebraska. On-base personnel were armed. Offutt became a haven for "Top Secret" clearances and the operations center for high-flying aircraft carrying atomic bombs. And SAC had a huge payroll. Good business for Storz Beer.

Todd, having no interest in the beer business, was attracted to the media mystique of radio; to the many signals and invisible voices he had heard through a crude receiver made from a cigar box, a crystal detector, a fine strand of wire, and an earphone that he had built when he was eight years old. That one gadget, called a "crystal set," is what hooked the imagination of young Todd to a radio career and the Storz name to a new brand of radio programming—Top 40.

Through his son Todd, the business presence of Robert H. Storz—the youngest surviving son of Gottlieb—would soon be felt in Omaha, and later in New Orleans, Kansas City, Minneapolis–St. Paul, Miami, Oklahoma City, and St. Louis. The reach and influence of Storz-style radio would become far bigger than that of Storz Beer.

Robert H. Storz, one of Gottlieb Storz's six children. In 1921, Robert H. Storz married Omaha native Mildred Todd. They had two children: Robert Todd Storz (the subject of this book) and a younger sister, Susan. To reduce confusion, the younger Robert Storz used his middle name "Todd."

By the time he turned 16, Todd Storz had expanded his knowledge about radio from a primitive crystal set to the status of a licensed "ham" amateur radio operator. He also took his chances as a 5-meter "bootlegger," operating without a license with teen friend Dale L. Moudy, who became a fellow "ham," and who would later be a contributor of essential engineering expertise in Todd's broadcasting ambitions.

Todd passed Federal Communications Commission "ham" license examinations for technical know-how, and the transmission of the dots and dashes of Morse Code telegraphy at up to thirteen words per minute. He set up his amateur radio station on the third floor of the family home in Omaha, proudly displaying his official FCC call letters, W9DYG.

Sister Susan recalls: "His radio room was directly above my bedroom. When he couldn't raise anyone with his call, he would ask me to come up and help. In my sexiest voice I would say, 'Calling CQ. Calling CQ' [ham-talk for 'Calling any amateur radio station']. Within seconds there would be many responses and I would be dismissed to go back to bed." This was a harbinger of her brother's later stunts to promote audience tune-in.

Perhaps because he was spending too much time tinkering with radio equipment, in 1940 Todd's father removed him from Omaha's Central High and enrolled him at the prestigious Choate School (now the co-ed Choate Rosemary Hall) near New Haven, Connecticut. For two years he was coached, tutored, mentored, and groomed for a college education by Choate's prep school "masters."

Interviewed by Richard W. Fatherley in 1965, Choate Alumni Director (and a former schoolmate of Todd's) Edward B. Ayres recalled: "Todd was some four years ahead of me, and because I knew very little about radio, I hardly saw Todd at all. He spent most of his time in the Choate radio room. Todd was not one of our best students because his tremendous love for radio conflicted with his studies. He had to learn the disciplines of study. He sometimes had to learn the hard way. Flagrant violation of the rules are not tolerated. However, he never registered any discontent or showed signs of wanting to leave. He was particularly frustrated in a couple of courses, but he didn't let this bother his grades. He always had this 'radio shack' to which he could retreat."

Ayres continued, "During Todd's first term of his first year [1940–1941], he was excused from athletics to work on his Class B radio license during his spare time. This was something he wanted to do. He didn't particularly want to compete in a varsity sport." Further, observed Ayres, "Todd was never considered a great athlete, probably because of his small stature. He was not a big boy by any means. However, he did select squash, but was not on the varsity.

"He did a good deal of work in forestry, an activity for boys who love the outdoors, and for boys who can't compete in contact sports," said Ayres.

Todd received two tongue-in-cheek accolades in Choate's 1942 class elections. He was described as "In Worst with the Faculty" and "Class Pest," sharing each "award" with several other classmates.

Choate's 1942 yearbook offered these captions under Todd's name: "Hot Toddy," "T.D.," "Todd," and "Sparks." At eighteen, he was described as 5 feet, 8 inches tall and weighing 130 pounds. His activities over the previous two years included squash, forestry, and being president of the Radio Club. His college of choice was the University of Nebraska. Under Todd's class pho-

Todd Storz in 1940, at the microphone in his "radio room" on the third floor of the family home in Omaha. On the wall are cards from some of the other amateur radio stations he had contacted. As can be seen, the call letters assigned to Todd's transmitter were "W9DYG." Fifteen years later, he would purchase his third AM radio station — WDGY in Minneapolis.

tograph was this epigram: "For, such as we are made of, such are we." Russell Ayres concluded, "I know nothing about the University of Nebraska where Todd is planning to attend, but I do know that Todd can make his stay there worthwhile if he wants to. I may be wrong, but I sense a great man in the making."

A contrasting view about Todd's future was offered by H. Dayton Niehaus, who taught physics at Choate. "Todd is planning to pursue a course in business administration at the University of Nebraska. Unless he gets in the army or navy service, he will not go beyond the amateur status in the radio world. I expect he will always have it as a hobby and will develop himself in it as he finds time to do more work."

After his graduation, Todd took examinations for the U.S. Army Cryptographers' School. In a letter to Todd's headmaster, Todd's father wrote: "He has attended several schools, the most recent being the Cryptographer School in the Signal Corps. I would hardly be a loyal father if I didn't tell you that he passed with the highest score of anyone applying for the Warrant Officer appointment."

Todd's friend and fellow ham, Roy Ekberg, became an FCC operator at that agency's radio monitoring station at Grand Island, Nebraska, working to

For, such as we are made of, such are we.

ROBERT TODD STORZ
"Hot Toddy" "T.D." "Todd" "Sparks"
Omaha, Nebraska

Age: 18   Height: 5 ft. 8 in.   Weight: 130 lbs.

Number of years in School: 2

1940-41: Tennis, Squash, Forestry, Radio Club
1941-42: Forestry, Squash, Forestry, President of the Radio Club

University of Nebraska

Closeup of a page from Todd's University of Nebraska yearbook. This photograph shows a more serious young man than the one in his "radio room" at home. Todd's slender stature is reflected in his height and weight numbers, while his love of radio is evident in his club membership.

monitor commercial broadcast stations, and also tuning in on wartime enemy communications as well as any illegal broadcasts. In 1942, the FCC received a tip. Said Eckberg, "A low power, illegal AM station was broadcasting somewhere within the University of Nebraska campus at Lincoln. I was asked to record sample transmissions for evidence. I learned later that Todd Storz had been that station's sponsor, and the FCC had shut down Todd's station."

Following a year at the University of Nebraska, Todd enlisted in the army in August 1943 and became a Warrant Officer Junior Grade in the Signal Corps. He achieved the highest score on the rigorous Signal Corps cryptography exams and became the youngest warrant officer in that branch of the service. He mustered out in April 1946. His service decorations and awards included the World War II Victory Medal, Meritorious Service Plaque, and the American Campaign Medal.

After being discharged, Todd enrolled in a 12-week radio course sponsored by the National Broadcasting Company (NBC) and Northwestern University at Evanston, Illinois. He subsequently took a job at KWBW in Hutchinson, Kansas, as an announcer, sales representative, and copywriter to gain some commercial radio experience. "I did everything there," Todd was later quoted as saying. "Engineering, announcing, selling, typing, writing copy, sweeping the floor." Multiple responsibilities were the norm in small-town radio stations.

He returned to Omaha in 1947 and joined KBON radio, a Mutual network affiliate. Todd hosted its *1490 Swing Club* from 11 P.M. to the station's midnight sign-off, where, according to station executive Paul Fry, he "employed a sort of Arthur Godfrey–Jack Parr style of sponsor-twitting." Engineer Don Mehl recalls Todd's instruction to a complaining listener — who didn't enjoy his selections of bop and swing music — to simply "switch their dial" and listen to somebody else. Todd soon moved into a sales position at KFAB in Omaha, and also started a reflective sign company, which quickly failed.

On November 1, 1947, Todd Storz married 23-year-old Elizabeth Ann Trailer at the First Presbyterian Church in Omaha. A glamorous honeymoon followed. "Mr. Storz and his bride will be at the Waldorf Astoria in New York City until they sail November 7 on the *Santa Paula* for a Caribbean cruise. They will go on to Caracas, Venezuela, and to the island of Curacao. They'll also visit Barranquilla and Cartagena, Colombia, before returning to Omaha," reported the *Omaha World Herald*.

When he resumed work at KFAB, Storz's duties included sales and marketing. While there, he became acquainted with Harold Soderlund, ten years his senior. Soderlund was a well-known radio program producer and advertising agency principal who handled the Storz Beer account. Over time, Soderlund became a confidant as well as an advisor to both Todd Storz and his father, as they approached their decision to purchase radio station KOWH from the *Omaha World Herald*—an acquisition that would soon dramatically change American radio programming. Under Storz ownership, KOWH would become the incubator and proving ground for the major elements of the Top 40 radio format, which in turn would have profound effects on radio in America for more than twenty years—between 1950 and the mid–1970s.

With the world watching what America did following the end of World War II, Americans began to watch television. Up to 1949, only ninety-seven television station construction permits had been issued by the FCC as the agency tried to work out numerous allocation issues. But once the "freeze" was lifted in mid–1952, 650 television station construction permits were soon issued. Back in 1950, only 9 percent of U.S. homes had a television set. By 1953, half of U.S. homes had one. The number of TV stations on the air climbed five-fold by 1957, to more than 550. By the end of 1959, 44 million American homes (86 percent) had television sets. The context for American radio in the postwar period now reflected the growing dominance of television.

Beginning around 1948, network television's first "season," network radio's most listened-to programs and their advertisers began defecting to "the tube," taking with them scores of radio's top programming and production people, sales and marketing personnel, technicians, and talent. Gone or going from network radio, were the likes of *Jack Benny, Our Miss Brooks, Dragnet, Superman, The Lone Ranger, Sky King, Gunsmoke,* and a laundry list of "soap operas." In Omaha, Nebraska, for example, young WOW radio morning show host Johnny Carson packed his bags with gags and left for the West Coast as a TV comedy writer, his first big step to becoming America's most celebrated television personality as the longtime host of *The Tonight Show* on NBC.

Meanwhile, local television sales and marketing teams were busy "switch-pitching" advertising agencies to shift their radio and print ad budgets to the newer television. While research about viewers was still rudimentary, it was clear that television was occupying more time in American homes every year—which had the effect of emptying local movie theaters of audiences, and radio stations of advertising revenues.

But television's ascendancy affected two types of radio stations very differently. From 1945 to 1962, network radio revenues declined from $134 million to their historic low of $44.9 million. With the blossoming of television, fueled in large part by the migration of many radio stars to the small screen, it was *network* radio that was quickly forced against the ropes. On the other hand, *local* radio revenues grew throughout the 1950s. Although stations affiliated with the major radio networks such as NBC, CBS, and Mutual had been the big money makers during the 1930s and 1940s, the situation changed dramatically in the decade of the 1950s. According to the 1958 study, "The Dynamic Change in Radio," prepared by Adam Young, Inc. (a national advertising sales representative for radio stations in the top 25 markets), the growth of local independent [non-network] radio over the network affiliate was "no longer a trend ... it's an accomplished fact." The Young report observed that until 1952, the most listened-to radio stations in the largest urban markets had been network affiliates. Just five years later, the leading radio stations in 21 of those markets were independents, demonstrating how local stations rapidly became the core of the radio business.

By 1968, local radio non-network "spot" advertising (advertising purchased in selected local markets rather than as a single nationwide network buy) totaled an unprecedented $1 billion. Although network radio had lost the living room to the television set, local radio stations had gained listeners in the kitchen, bedroom, bathroom, den, backyard, and on the road. The growth in car radios in the 20-year span between 1945 and 1965 was also dramatic, rising from just 23 percent in 1945, to 60 percent in 1955, and to 79 percent in 1965, when over 14 million vehicles had radios. By 1957, one-third of all U.S. radio listening was taking place in cars. As television sets supplanted radio receivers in the living room, the availability of listeners in cars at any time of the day became increasingly important to the radio industry. With the growth of suburban communities ringing major cities in the 1950s and 1960s, increasing numbers of listeners in cars commuting to and from work were identified as "drive-time" audiences. This valuable "captive" audience could not be reached by television or print advertising. In the evening, while mom and dad were home watching television, teenagers were driving around for the sake of driving around, listening to radio stations that were programmed specifically to appeal to young adults.

Changes in the radio receivers themselves also influenced radio listening patterns. The first transistor radios were sold in 1954. By the early 1960s, transistors had for the most part replaced tubes in radio receivers. Radios became less expensive and more portable, so additional radios were bought for the kitchen and the bedroom. As a result, individuals listened to radio almost anywhere *except* the living room, where the television set had supplanted the large old tube-type radio as the provider of family entertainment. A self-promotional phrase used by the radio industry at the time was "the smaller radios get, the bigger radio gets." It was inarguable. The diaspora of radio sets to multiple rooms of the home and into automobiles allowed listening to happen almost anywhere and anytime. As a result, radio revenues — which had nosedived during the explosion of television viewing back in the 1950s — began to rise again in the early-to-mid–1960s. As that decade came to a close, the revenues realized by many music-oriented AM radio stations peaked, and FM radio broadcasters at last began to realize profits from growing audiences attracted by the fidelity of FM receivers and a wider choice of musical genres.

Just as the advent of television in American homes in the early 1950s furnished the "left-hand" bookend of the Top 40 era, the ascendancy of FM pop music stations in the early 1970s would become the "right-hand" bookend to the story of a group of six Storz-owned "music-and-news" stations that became the proving ground for radio's Top 40 revolution.

• Two •

The Incubator: KOWH, Omaha

In mid–April of 1949, Todd and Robert Storz announced their purchase of KOWH and its FM companion outlet, KOAD, from the World Publishing Company, owners of the *Omaha World Herald*.

On the last Wednesday of the month, they motored the sixty miles south to Lincoln, the state's capitol, to file their papers with the State Corporation Commission to form the Mid-Continent Broadcasting Company, the intended licensee of the radio station (subject to FCC approval). The *Omaha World Herald* reported, "The concern was capitalized at $250,000 by Robert H. Storz and Todd Storz as incorporators," after they purchased the properties for a total of $75,000 (equal to about $714,248 in 2012).

Under the newspaper's ownership, KOWH "wasn't making much [money], and the *World Herald* had tired of the whole idea," reported the *Omaha Sun*; hence the unexpected opportunity to buy KOWH at a bargain price. Todd chipped in $20,000 from a mortgage on an Iowa farm, his father contributed $30,000, and another $25,000 was borrowed from a bank. At last, Todd would have a real commercial radio station to operate.

In mid–July 1949, the FCC approved the transfer to Storz. "The stations will become the property of the Mid-Continent Broadcasting Company. Robert H. Storz is president of the new company. His son Todd will be vice-president and general manager," noted the *Omaha Sun*.

The KOWH studios were located on the eighth floor of the Kilpatrick Building at Fifteenth and Farnam streets in downtown Omaha. The KOWH 40-acre transmitter site was six miles northwest of Omaha near the suburb of Benson. KOWH was not a prime facility, for two significant reasons. First, it was a daytime-only station, meaning that it was required to sign off at sunset. (After sunset, colder air in the upper atmosphere can reflect radio signals multiple times, causing interference to distant stations on the same frequency.) Second, KOWH's power was capped at just 500 watts, compared to the 50,000-watt output of the most powerful AM stations. KOWH's power was limited to 500 watts to keep it from interfering with the signal of WEAF in New York City, which transmitted on the same 660 kHz frequency as KOWH. The FCC had designated WEAF as a "clear channel" station — one that could be heard without interference up and down the east coast of America. (WEAF's later call letters included WNBC, WRCA, and WFAN.)

KOWH's $75,000 purchase price included its sister station, KOAD-FM. In 1946, KOAD had become the first FM radio outlet in Nebraska. But just three years later, Storz returned the license to the FCC, because with only a few in-home and in-car FM receivers in the Omaha area, keeping the station on the air was a waste of money. In the words of KOWH announcer George W. "Bud" Armstrong, FM in those days stood for "Failing Money." As will be shown later, KOAD-FM's inability to show even a single profitable month under Storz ownership contributed to Robert H. Storz's antipathy to FM broadcasting. The KOAD-FM frequency was transferred to the University of Omaha (now the University of Nebraska–Omaha).

Although KOWH was limited to daytime-only operation, its relatively low 660 kHz position on the AM radio band allowed its signal to reach well into east-central Nebraska, west-central Iowa, northwest Missouri, and northeast Kansas. AM radio stations with frequencies at the low-numbered end of the frequency spectrum have much broader signal coverage than AM stations at the high end, and require less electricity to operate in the bargain. The "longer" waves at low frequencies simply spread out farther than the shorter waves at higher frequencies. Locally, KOWH's signal strength was comparable to other Omaha stations.

One of Todd Storz's first decisions as the new licensee was to maximize the loudness of the KOWH signal at the transmitter site, as he had done with his ham radio. That was accomplished by adding special "limiting" amplifiers that never allowed the audio level being sent to the transmitter to drop below a certain loudness threshold. "Todd had the loudest signal on the air!" exclaimed fellow ham Don Mehl.

The FCC's mandate that daytime-only stations such as KOWH must sign off at sunset caused "daytimers" to be less valuable than full-time facilities. It is one of the reasons why Storz was able to buy KOWH for such a relatively low price. But in a fine example of "turning lemons into lemonade," Storz decided to refurbish the KOWH studios so that they would be a first-class production facility for use during the nighttime hours, when the station was off the air.

The choices made in equipping those studios provide a glimpse of the rapid technological changes in sound recording equipment at that time. In the early 1950s, reel-to-reel tape recording equipment was a work-in-progress, based on World War II recording technology that had been captured from the Germans just a half decade earlier. The chemistry of getting magnetizable iron oxide particles to stick faithfully to flexible strips of paper or plastic tape was still unsettled. Fidelity was lacking in the high frequency range where consonants made words understandable. There was not even agreement on whether recording tape should travel from left-to-right or right-to-left. Because Storz's recording needs were immediate, professional-grade recordings meeting the demands of studio and broadcast quality standards were not recorded on tape, but instead were recorded on "acetates"— the nickname for the discs which were produced the same way that the original "master" discs of music recordings had been made for decades. The KOWH recording studio employed the reliable disc-lathe equipment which record companies used, with a stylus cutting a continuous circular groove of audio onto a black lacquer-coated aluminum disc. The finished "acetate" discs were like a larger, heavier version of the LP record, which had appeared in the late 1940s.

According to engineer Dale Moudy, who developed cost estimates for the recording studio (which had to be approved by Todd's father, Robert H. Storz), a professional disc-cutting lathe could cost up to $2,500 (about $23,334 in 2012).

The production studio turned out to be a good investment. At night, the KOWH studios became a recording center for Omaha advertising agencies, and for local advertisers whose commercials would be broadcast on KOWH and other stations. Because the studios offered superior audio quality, they also became a profit center for the station, while projecting an image of professionalism. In publicity about the facility, Dale Moudy boasted that "KOWH also does a considerable amount of commercial recording, the majority of which are spot announcements produced by the program department. Perhaps some have been used in your city!"

One of KOWH's "signature" voices in the production studio — as well as on the air — was that of Don Loughnane. His on-air sound matched his outsized six foot four inch frame. At KOWH, Loughnane was an attention-getting news reader, but he also recorded news introductions, sponsor credits, contests, promotional announcements and more in his recognizable and believable voice. Via those recordings, his voice would become the signature sound of later Storz station acquisitions in New Orleans, Kansas City, and Minneapolis–St. Paul.

In regard to KOWH's own programming in 1950 — a year after Storz acquired the license — it remained a patchwork of local programs, network shows, and even syndicated British soap opera recordings. After six months of losses, KOWH did manage to show a year-end profit of $84 (about $784 in 2012).

World Publishing Company, the previous licensee, had left behind a core of employees ready for Storz ownership. These included staff announcers George W. "Bud" Armstrong and James O'Neill; sportscaster and salesman Jack Sandler; record librarian Margaret E. "Peggy" McGrath; and technical director (engineer) Dale L. Moudy.

The names are of more than passing interest. All of these early KOWH colleagues continued with Storz through five station acquisitions in the decades that followed. Bud Armstrong succeeded Todd Storz as head of the company when Storz died unexpectedly.

In 1965, Richard W. Fatherley interviewed Dale Moudy about the earliest days of KOWH under Storz ownership:

> When Todd took over in July of 1949, he wanted to get things done. I recall a contest he had among the station's personnel. The person who submitted the most workable idea would win a $100 War Bond [about $945 in 2012]. I won that contest by proposing an hourly [on-air] time-tone keyed to the self-setting Western Union clock. [At the precise second of the new hour, the Western Union Telegraph Company sent an electric pulse to their special clocks which reset them to exactly "the top of the hour"— Moudy's invention turned that internal signal into an audible time tone.] "It was a simple thing, but a great advance for independent radio at the time. In addition, I proposed rebuilding the control room to improve the operation of the station in ... programming and production. In February 1950, I was promoted to studio supervisor by Todd. He wanted a lot of things done that today aren't considered great, but at that time were considered vital; little things that helped the sound of the station.
>
> Another thing we did at KOWH was putting Santa Claus on the air answering the telephone from the North Pole. We talked about it one morning and Todd wanted to have something that would answer the telephone, play a record[ing], and hang up. This machine that we developed was the forerunner of the answering device that's in use today.

Moudy recalls that he ordered the parts to put the device together, sketched out a circuit, and built the machine by late that afternoon:

> The telephone company said we couldn't run the promotion because we were planning on attaching "foreign" equipment [circuitry that the telephone company didn't build and license] to their lines. We then asked when they planned to make available a similar device. They said it would take four years to develop. Shortly afterward, the telephone answering device did make its debut. However, to illustrate my point, we did the job in one day.
>
> Todd believed in developing an idea, spending the money, and finishing the

job as fast as possible. We built transmitters, consoles, and other equipment simply because there wasn't anything available that could meet our needs. Todd was gadget-minded.

Family friend and advisor Harold Soderlund agreed. "Yes, he [Todd] was an engineer at heart. He truly was. He was interested in the engineering — in the sound."

KOWH's changeover to Storz management turned out to be a positive experience for KOWH's veterans. "After all the apprehension, speculation, and second-guessing on our part, the course of events after the Storz takeover was anti-climactic," said announcer James O'Neill in his memoirs. "The station got new furniture in the lobby; there was repainting; but for personnel, no bloodbath ensued. We got a new program director, some additional sales people were hired, and the light concert music was discarded," he said.

Storz hired short, dark and handsome Gaylord Avery to be his new program director. Avery despised television. He had been a news reader at Omaha's KFAB radio, and was an often-requested commercial voice. His mellifluous pipe organ baritone delivery later became his ticket to CBS in New York City as a television booth announcer. Among Avery's first assignments at KOWH was accompanying Storz to New York City to call on leading advertising agencies, and to monitor the programming of WNEW, an independent station that programmed popular music recordings hosted by strong air personalities, including Martin Block on his *Make Believe Ballroom* show. Block was gifted with a listenable voice, believability, clear diction, good pacing, and knowledge about popular music and its personalities. Storz understood that these attributes helped to shape the "air face" of a successful radio station, and thus he sought articulate, big voices like Block's as candidates for air staff positions on KOWH.

KOWH staff announcer O'Neill admitted in his memoirs that he could not recall what the station did with music in the first few months under Storz ownership (in 1950):

> What I do remember is an eclectic sound. At sign-on [6 A.M.], we had the old "Sunshine Maker." That's what he had painted on his car. His name was Sterling Welker, a nice man who came on like an old-time tent-show barker. He took requests, offered homey tips, and had a loyal following. Sterling was followed by one of our salesmen, a man with on-air experience who played the ukulele. We continued to broadcast *Kitchen Klatter*, a daily hour from Leanna Drifmeyer's kitchen in Shenandoah, Iowa. Todd bought two hourly shows on a daily basis. They were British Broadcasting Corporation [BBC] productions: *Gommards of the Courts* and *Great Moments In Medicine*. We had sports news and interview shows with Jack Sandler, and a public library storytelling section for little nippers. I assumed an old man's voice, for all my 21 years: on Saturday I became

Uncle Phineas with children's story records. On summer nights when [as a daytime-only station] we could stay on until 8:30, I did a country-western gig. The point of all of this is, I'm not sure Todd yet [in the very early 1950s] knew what he wanted from his radio station.

Former KFAB sales manager and Storz confidant Harold Soderland put it this way: "Todd stumbled around with KOWH for the first year. It took him a year to really find out where he was going to go with this thing." That assessment was echoed by Dale Moudy in his 1965 interview with Richard W. Fatherley: "When Todd bought KOWH he didn't know what to do with it.... We had a conglomeration of programming on the air.... We tried fighting the networks with canned soap operas that we bought from England. We tried hour blocks of news that didn't accomplish anything." Indeed, a very basic local radio station ratings service showed that in 1949, only four out of 100 Omaha homes listened any time at all to KOWH.

By the spring of 1951, the beginning of the transition to a more music-oriented format was taking shape. A 2003 interview by Tom McCourt with KOWH's record librarian Peggy McGrath yielded the following KOWH program lineup as of March 1951:

6:00 A.M.—*Farm Hand and News*
7:00 A.M.—*Musical Clock, News*
8:00 A.M.—*Kolache Club*[1]
9:00 A.M.—*Hatchery Program*
10:00 A.M.—*Back to the Bible*
10:30 A.M.—*Today's Top Tunes*
11:00 A.M.—*Kitchen Klatter*
11:30 A.M.—*Lawrence Welk*
12:00 noon—*Barn Dance*
12:15 P.M.—*Kolache Club*
1:00 P.M.—*Make Believe Ballroom*
2:00 P.M.—*The Sandy Jackson Show*
4:00 P.M.—*Sweet Music*
4:30 P.M.—*The Jim O'Neill Show*
5:00 P.M.—*News and Weather*
5:30 P.M.—*Easy Rhythm*
6:00 P.M.—*Sports Trail*
6:30 P.M.—*Eventide*
6:45 P.M.—*Air Force Show*
7:00 P.M.—*Today's Top Tunes*
7:30 P.M.—*Signoff*

As can be seen in the program schedule, music programming dominated KOWH's afternoon hours in March of 1951, but mornings consisted largely of syndicated, pre-recorded, and non-local programs. Most "independent" (non–network-affiliated) stations such as KOWH aired a mix of recorded and live programs from a variety of sources. One of the few "live" programs that came to KOWH by telephone line was *Kitchen Klatter*, featuring women offering recipes, discussing horticulture, debating family matters, and answering household questions. "I think this program was the tip-off because our ratings went down," recalled Sharpe. "The audience came back almost two hours later, when music programming returned," he noted. Peggy McGrath concurred, saying, "The thing I remember most were the adjacencies. [Storz] figured out 'We're not getting any audience next to [*Kitchen Klatter* host] Martha, she's got to go.'" Even though it was a sponsored program that for years had been profitable for KOWH, *Kitchen Klatter* was canceled and replaced with popular music.

It is important to understand why it was appropriate to cancel *Kitchen Klatter* (and other programs like it). In the beginning years under Storz ownership, KOWH aired programs that produced revenue through an agreement with the program provider to buy the station's airtime to run the show and the provider's commercials. Most independent stations operated on that revenue model: the station's transmitter was for hire, and the program provider paid the bills. The problem was, program content was effectively in the hands of the advertiser. Consistency of program content was nearly impossible, and quality was variable. Storz decided to abandon this old "airtime-for-sale" business and instead adopt an "audience-for-sale" model. What advertisers would be offered was a growing audience of listeners to KOWH's own internally generated programming, as measured at regular intervals by C.E. Hooper's ratings system. Producing programming that could attract and hold audiences henceforth would be 100 percent the responsibility of KOWH's own programming people. The air talent had to be more than merely mellifluous announcers, or passive babysitters of the record turntables. They had to be personable local entertainers.

According to McCourt and Rothenbuhler, *Kitchen Klatter* was the last outside program to be dropped. It left KOWH's air at the end of August 1953. They affirmed that "by September of 1953, all block programming was gone from the station." From then on, KOWH's programming was all produced "in-house" and most of the ten or so locally produced shows featured recorded popular music and local announcers.

One such show stood out in the recollections of both Dale Moudy and Virgil Sharpe. Said Moudy: "The key was a show called *Sweet Music*, when we played top music of the day. It was, more or less, the hits of the day, originally hosted by Bud Armstrong." The show had begun airing in March of 1950, in the 4 P.M. time slot. It became very popular. One year later, that program was surrounded by other music programs that stretched from noon to KOWH's sign-off at sunset. Dale Moudy remembered that "one day Todd was approached by [an administrator from] the University of Nebraska [at Omaha]. They stated they had conducted a survey and asked if Todd would like to buy it. Naturally, he was curious and bought the thing. Amazing as it seems, the survey ... showed the *Sweet Music* show had the largest share of audience of any other time of our day. So, Todd expanded the program and the Hooper Ratings confirmed its popularity."

Virgil Sharpe corroborates Dale Moudy's account of discovering that the 4 P.M. *Sweet Music* show had the largest share of audience: "One of the all-music programs on KOWH seemed to be getting a higher rating than some of our other programs during daylight hours. And the air personality carrying the program seemed to pick up and lift the ratings as well." The increasing tune-in to the *Sweet Music* program in 1951 allowed KOWH to boast that its ratings had increased 600 percent compared to 1949. But that surprising number reflected the station's paltry listenership in 1949 at least as much as its growing audiences in 1951.

Virgil Shape's assessment of why Storz bought the University of Nebraska–Omaha audience survey suggests an interest in more than merely acquiring a tool to sell airtime:

> He was interested in what people listened to and why. What motivates them to listen? What he was after was: What do men listen to and what do women listen to? The survey revealed music had a high appeal to most everyone.... The results were compiled with the answers being: Music was an important item; an essential ingredient in the motivation of listenership in radio. So, whenever we programmed blocks of music, the interest would go up.

Storz made two crucial decisions which set KOWH on the path to success. One was that he sought out objective third-party data about program popularity, instead of assuming that the assortment of different shows KOWH carried would offer "something for everybody." Second, based on that objective data, Storz made the decision to expand the availability of the most popular programming, in effect "super-serving" a narrow but growing segment of the audience. In an era when network radio, local radio, and emerging television stations were all offering a growing variety of short programs to appeal to a certain slice of the available audience for a finite amount of time, Storz began to offer one program type — recorded popular music — that soon would expand to be heard in all day parts.

To begin to put the new hit records programming on the air, Storz hired disc jockey[2] Sandy Jackson from KBON, where Storz had worked in sales. Jackson had started playing records requested by listeners back in 1944, on *The 1490 Swing Club* show which aired between 11 P.M. and midnight on KBON — the program that Todd Storz had hosted back in 1947 when he returned to Omaha after leaving the station in Hutchinson, Kansas. As one of the city's top radio personalities, Jackson's exodus from KBON to join KOWH was a big event in Omaha radio. "It was a major coup when we got him," remembered Peggy McGrath. Jackson had what listeners described as a friendly, melodic voice. He played an interesting blend of popular music on his KOWH afternoon show, and he invited listeners to mail him their record requests and song "dedications" (e.g., "This one goes out to Danny, from Sue at the Elmore Bakery...") Jackson made on-air dedications in exchange for a twenty-five-cent (about $2.16 in 2012) donation to "The Welfare Fund." In those early days of the Storz operation, he was the only KOWH air personality who was allowed to do an all-request show. Listeners who recall him say he didn't pander to his audience, but seemed to have multiple ways of pleasing them. Some described his presentation as "magic." Unfortunately, a full recording of Jackson on KOWH in that era has never been located, but a brief clip of a KOWH newscast, Jackson's theme song, and his show introduction was discovered by Deane Johnson.[3]

Sandy Jackson's runaway success on KOWH resulted in audience numbers for an independent station that hadn't been seen before in the Omaha market, and which were even rare nationally. Jackson consistently drew Hooper ratings in excess of half of all Omaha listeners.[4]

Next, Storz brought in Iowa disc jockey Johnny Pearson to do the morning show on KOWH. Pearson's dry wit and falsetto-voiced alter ego "Amanda" stole morning audiences from none other than Johnny Carson on Omaha's powerful WOW. (Carson would later head for Hollywood to become one of America's all-time favorite television celebrities.) A memorable stunt during Pearson's early tenure on KOWH occurred on a cold December morning when he announced a $25 cash prize to the first woman who'd meet him at the corner of Sixteenth and Farnam in a bikini with a rose in her teeth. That was a lot of pocket money in 1952 (about $212 in 2012), and yes, a woman did show up in a bikini and with a rose in her teeth. Later, Pearson would move to Storz's WHB in Kansas City where he again obliterated the competition in his time slot.

During the early days of KOWH, Storz made certain the week's top 10 songs as determined by the popular *Your Hit Parade* network program were played using the original versions by the original artists on 78 rpm records. As a result, KOWH ratings nudged farther upwards.

Sandy Jackson. Sandy Jackson was Todd Storz's "voice of choice" to attract listeners to KOWH's then-radical tactic of playing recorded popular music exclusively. Jackson had attracted large audiences to his program on Omaha's KBON, where Storz had previously worked in sales. Getting him to move to KOWH was described as "a major coup." Jackson had a friendly, melodic voice and was happy to take song requests.

In *This Business of Radio Programming*, Claude Hall, former radio editor for *Billboard* magazine's "VOX JOX" column, reproduced an interview with respected program director Chuck Blore that confirmed the importance to KOWH of playing the top 10 hits. Said Blore: "Todd Storz's ... initial concept was in playing 10 records. That was it, initially." Further, said Blore, "the information he got initially was from jukebox operators, and they were telling him which records were being played most. He took the top 10 records and played them over and over and over again."

Perhaps that last sentence is an overstatement. The top 10 records did become a "must-play" in each music program, but at first they were bracketed by popular standards from the KOWH music library. The record-choosing process was supervised by Peggy McGrath, who kept tabs on current hits in *Billboard* magazine, *Cashbox*, and *Variety*. She also maintained a log of requests from Sandy Jackson's show, since Jackson specialized in requests.

Later, in addition to tracking Omaha record store sales and jukebox plays, KOWH built its own weekly list of current most-requested record titles. There were no "oldies" played in the early days of the format.

KOWH's programming experiments had shown that playing popular music, and repeating the top-selling hits more often, were the main engines driving KOWH's audience growth. But in the 1950s, the Omaha market wasn't interested only in escapist entertainment. The Cold War was very much on the minds of many local residents, and especially the air force personnel assigned to the Strategic Air Command headquarters at Offutt Air Force Base. Combat intelligence of interest to the personnel at Offut sometimes found its way into the KOWH newsroom not only via several wire services, but also from Storz-friendly contacts on-base.

Beginning in 1951, Storz scheduled KOWH news at five minutes before the hour under the banner of KOWH News "Live at :55," a news concept adopted seven years later by the ABC Radio Network. Storz's "Live at :55" idea promised that KOWH would air essential news before the network newscasts at the top of the hour. "Live at :55" also allowed KOWH to restart its music and fun programming just when other local stations were beginning their hourly network news. "Live at :55" news allowed KOWH to "scoop" Omaha's network-affiliated stations by five minutes, and to program popular music at "the top of the hour" for listeners avoiding network newscasts.

On June 15, 1951, the *Omaha World Herald* broke some very big radio news in a tiny one-inch article titled "KOWH Ranks High." The article said that "KOWH of Omaha was rated first in the United States for having the largest percentage of audience of any independent radio station in the country." The word "independent" was the important qualifier, because network affiliates were still generating large numbers. The brief article gave a nod to the C.E. Hooper rating company, which had pointed out that KOWH led independent stations in much larger markets such as Baltimore, New York, Kansas City, and Chicago. KOWH was ranked first among 145 stations across all large cities.

Nebraska governor Val Peterson issued a congratulatory statement, saying, "I think it is grand that a Nebraska station, for the first time in the history of the state, has been rated as having the largest percentage of audience of any independent station in the country. It's a wonderful honor for your station and for Nebraska."

Two years later, that same claim — of having the largest share of audience of any independent station in America — was repeated in a KOWH advertisement in *Broadcasting-Telecasting* magazine on October 12, 1953. The station's success had certainly not been a mere "flash in the pan."

It was inevitable that KOWH's success in building audiences began to attract would-be imitators. But Storz programming had not been reduced to a universal formula for listener fun and station profit which could be easily cloned. In fact, when the term "formula radio" was later used as a pejorative term by Storz's detractors, the company rightly denied that their programming could be reproduced via something as simple and rigid as a formula. As Storz began to acquire stations in other markets — initially WTIX in New Orleans, then WHB in Kansas City, and a bit later WDGY in Minneapolis–St. Paul — each would have most of the elements that had been developed at KOWH, but each would also have its own slightly specialized local sound. That "localization" took into account the makeup of the available audience and the programming on competing stations. Broadcasters who tried to clone the Storz sound often assumed that they merely needed to figure out a singular "formula," but the basic recipe was altered to match each market. It was the difference between buying a suit off the rack or having one sized and sewn by a tailor: tailoring produced a better fit.

By 1951, Storz was in his second year as owner of KOWH. His air staff and promotional ideas were attracting attention from listeners and local media including newspapers, wire services, and the new Omaha television stations which enjoyed jabbing KOWH as an upstart. Live television programming in America's heartland was almost entirely local, because the expensive "coaxial" cables that offered enough bandwidth to carry TV signals were mostly concentrated on the east coast where the NBC, CBS and ABC television networks originated. But in August of 1951, AT&T's Long Lines division formally opened its new microwave relay system. The telephone company's elaborate steel towers with odd-shaped "horns" for receiving and retransmitting extremely high-frequency microwave signals were soon being built across America, to replace many of the coaxial cables. Thanks to the new microwave relay towers, ABC, CBS, and NBC television programming would become available to homes nationwide. Network television was grabbing all the headlines, whereas both local and network radio stations were being threatened with the loss of listeners (which would translate into a loss of advertising dollars). Showing growth in radio listening had become a key to survival.

KOWH listenership continued to be measured by the C.E. Hooper, Inc., radio ratings service, whose employees made monthly telephone calls in the Omaha region to determine radio listening trends. It was obvious to Todd Storz that the Hooper ratings verification of growing listenership to KOWH was a key to increasing radio advertising sales. Said Storz in his KOWH sales brochure *What Makes KOWH of Omaha Tick?:* "The C.E. Hooper Company has gained national prominence and acceptance ... because of its reliability and integrity. Never in the history of the Hooper Company ... has any survey been proved false or fraudulent." As Storz's new programming attracted larger

Todd Storz in 1956 — at work in his office at KOWH. He was a very busy man by then, having acquired WTIX in New Orleans, WHB in Kansas City, WDGY in Minneapolis, and negotiating to buy WQAM in Miami. Notice the multi-band ("shortwave") radio receiver to Todd's left.

audiences, competing radio stations often cried "foul" by lashing out at the ratings companies who delivered the (to them) bad news. Storz's defense against charges leveled by competitors — i.e., that the Hooper ratings were "fraudulent" — was wholehearted and effective. He would use the Hooper ratings service with each additional radio station acquisition, remaining a loyal subscriber. (Later, the A. C. Nielsen ratings company took over Hooper.)

Storz understood that non-network affiliated stations such as KOWH had to work harder than the network affiliates to attract tune-in. Network programs received attention in national newspapers and magazines, but locally programmed stations had to continuously develop and deploy their own audience promotions. Offering innovative and more interesting programming created loyal listeners, but in order to grow station audiences, it was necessary to recruit and retain people who had never tuned in. Good programming kept people listening, but it would be station promotions that caused non-listeners to give the station a try. Thus, attention grabbing promotions became an essential part of KOWH's programming and would be cloned at the other stations Storz acquired.

The Storz promotions which garnered the most attention in the early days centered around the lure of "free money." The transfer of that money from the station to the public could be as simple as circulating $1 bills with an attached note that said, "Have lunch on us!" Early KOWH announcer James O'Neill recalled, "In the very early 1950s a dollar [about $8.65 in 2012] would buy lunch. Not at a five-star restaurant, but lunch." It was a clever promotion that was sent to a list of present and potential advertisers.

On October 22, 1951, the *Omaha World Herald* covered Storz's first treasure hunt — which actually was rather tame compared to iterations of the same stunt in subsequent years. The newspaper merely reported that it caused traffic jams "inspired by clues supplied by KOWH," leading to the whereabouts of money "planted at various points in the Omaha and Council Bluffs area." While driving with treasure hunters on streets crowded bumper-to-bumper to check on the progress of the promotion, Storz tried to exit the jam by turning onto a side street. He was arrested for "failure to stay in the line of traffic." An officer demanded of Storz: "Do you want to go to jail?" Instead, Storz was released on $10 bond.

Ignoring the adage that "money doesn't grow on trees," another of Storz's early and very successful promotions tried to make it seem that money did grow on trees for listeners to KOWH. In late February 1952, air personality James O'Neill threw $1, $5, and $10 bills from a tree in Omaha's Turner Park. Storz arranged for O'Neill's sudden departure from his regular program, which was interrupted with an announcement that the safe door in the KOWH business office had been found "open," and the office wastepaper basket was "missing." After explaining the circumstances, O'Neill's stand-in announcer said an unnamed physician had declared that O'Neill had "flipped out" and was throwing away money while sitting up in a tree at the park. In other words, the station broadcast false "news" as a sheer stunt.

In his memoirs, O'Neill remembered: "Arms from a growing crowd reached higher. What's that drawing up to the curb? Two police squad cars. Whatever can they want? They want me out of the tree. I comply. One cop pulls my arms behind me and snaps on the cuffs. Nobody read me my rights. The Miranda incident is far in the future."

Charged with disturbing the peace, inciting a riot, and causing snarled traffic, O'Neill's bond was set at $100 ($865 in 2012). The remaining cash was seized by police as evidence. Meanwhile, keeping up the stunt, KOWH asked the public to go to the police station to bail O'Neill out, tying up traffic once again. The police were not amused and handed out tickets to those who were double-parked.

Said O'Neill, "Todd Storz firmly believed in surprise and audacity as audience-building tools." The stunt had been Storz's idea — every minute and every nuance of it. He would develop elaborations of that promotion for KOWH and for the additional stations he would acquire. Later versions would offer much bigger jackpots, resulting in huge traffic jams, occasional property destruction, and sometimes hefty fines levied against the station.

Why take the risk? Showing growth in listenership was the key to attracting advertising dollars — at first locally, and then nationally. A brochure sent to potential KOWH advertisers from that period put it this way: "We pledge to you, our advertisers, that every program policy will be geared to obtain a bigger volume of listeners to KOWH and to your commercial message."

Although money giveaways tended to garner the headlines, simpler and less costly promotions were equally important at KOWH because they gave the station a sort of "lighthearted personality." Helen Stacy Norwood was hired by Storz from Omaha's KOIL to handle the KOWH continuity department. "He could fill [i.e., perform] any position on his staff," Helen recalled. "The one exception was copy writing. Todd readily admitted that he could not fill this job. And that's where I came in."

Norwood supplied a glimpse of how promotional ideas (such as throwing money from a tree) came into being, in her short story "A Tweety Bird Christmas":

> Storz was unique with his staff, treating them as peers rather than employees. Each morning around 10 o'clock, department heads and anyone who wasn't on the air at the time went to Bishop's Cafeteria for coffee. Beaucoup ideas were

born at these klatches, and plans put into effect, encouraged and led by Storz. As the Christmas season approached, a unique plan was offered for listener participation. The first person to teach a pet bird to give a station break would receive $25 [about $212 in 2012]. The phrase must be repeated distinctly and be recorded [at the KOWH studios] in order to be eligible to win. Days passed with no response. Then, on the fifth day an excited caller told us her parakeet could indeed say the words perfectly: "Merry Christmas, KOWH, Omaha, Nebraska."

The lady and her bird were assigned to a recording room. The microphone was turned on and the door closed. For two full hours our contestant tried in vain to get her beautiful feathered friend to perform. We suggested she take a break for lunch and leave the parakeet alone in the small studio. I decided to turn on the speaker to capture the words. My startled ears heard, loud and clear, over and over, "Merry Christmas, KOWH, Omaha, Nebraska. Here comes that darn paper boy. Merry Christmas!"

On a sunny day in April 1953, KOWH rebroadcast the emergency announcements that were part of the previous year's coverage of Missouri River flooding. Dozens of men arrived at the city hall to volunteer for the presumed disaster. The local nurses' association wanted to know why they hadn't been notified, and the mayor received telephone calls at his office. Switchboards to local authorities were jammed, and many people left their jobs to return home expecting the worst, reported the *Omaha World Herald*. KOWH said the rebroadcast was made to "keep people awake to the ever-present threat of an emergency." It was a flimsy excuse by the station, but the newspaper account admitted that, "the 25-minute program ... was preceded and ended by announcements that it was a rebroadcast."

In July 1953, Storz rattled some cages within the Omaha establishment. Armed with a hidden (though book-size) wire recorder attached to a Dick Tracy "style wristwatch microphone, KOWH newsman Don Loughnane captured the sounds of illegal after-hours whiskey being poured, the rolling of unlawful dice, and the remarks of several prominent figures allegedly participating in behind-closed-doors gambling. According to the *Omaha World Herald* of July 10, 1953, Storz had paid $500 (about $4,212 in 2012) for the German-made recording device, and bankrolled Loughnane's surreptitious bar-hopping recordings with $700 (about $5,897 in 2012) for food, drinks, and gambling. Portions of the recordings were played on Loughnane's "Omaha After Dark" broadcast. The newspaper reported that KOWH had five additional recordings that were not broadcast because — in Storz's words —"some were ruined by profanity, and one involving some prominent figures were believed to be 'too hot' to release." There were demands by the Nebraska Liquor Control Commission, the state Bureau of Internal Revenue, and the Douglas County attorney for information on the broadcast, but Storz declined to cooperate without a court order.

One of KOWH's more outrageous promotions placed checks in selected books at the Omaha Public Library to — as KOWH put it — "encourage better patronage of the library." The stunt caused pandemonium in the library's quiet, staid readings rooms. Some 90 volumes were damaged or destroyed. Eventually, KOWH paid $565 (about $4,760 in 2012) in book replacement costs and damages.

While these Storz promotions may have failed to be in the "public interest, convenience, or necessity" (a benchmark used by the Federal Communications Commission in evaluating the performance of radio station licensees), they did succeed in igniting controversy and conversation as ever-larger numbers of people began to sample the station's programming — and that was the whole point.

KOWH's then-program director Virgil Sharpe put it this way: "All of this caused a great deal of attention. But this was what KOWH needed! We had what we thought was a good [program] product, but we had to draw attention to it first, just like the old story of hitting the mule on its head with a 2-by-4 to get its attention. Todd knew this — either instinctively, or he found out about it through his own experience — but he knew it."

Several other important business concepts began to take shape at about this time, to be further developed and formalized in the following years. The first significant departure from standard radio business practices was that Todd Storz was not interested in selling radio program time, radio commercial "spot" announcements, or any other form of radio advertising merely for the sake of revenue. The Storz radio business model comprised three steps: (1) create a programming product that attracts audience; (2) measure that audience with reliable survey methods; and then (3) sell that audience to advertisers. In short, Storz stopped being in the business of selling "time" on the station's airwaves. Every radio station had time to sell. Instead, Storz was in the business of selling audience — in the case of KOWH, a large, loyal, and continuously growing audience. In his KOWH brochure titled *What Makes KOWH of Omaha Tick?*, he spelled it out this way:

> We know of only one way to get results. That is to have an audience in large enough numbers for every one of your commercials and programs. Just as the regular salesman on the street, your radio salesman "MUST MAKE CALLS TO GET RESULTS!" Therefore, we pledge to you, our advertisers, that every program policy will be geared to obtain a bigger volume of listeners to KOWH and to your commercial message.

This commitment to build large audiences — "to obtain a bigger volume of listeners to KOWH and to your commercial message" — turned out to be an expensive undertaking. In the early 1950s, with TV stations gradually expanding their program schedules from the evening into daytime hours — thereby further draining audience from radio — the temptation was to keep radio's operating costs (such as salaries, programming, and promotion) as low as possible. However, Storz decided that he was mistaken in trying to bring overhead below revenue, instead of trying to raise the revenue. Because salaries were a substantial portion of operating expenses, keeping them as low as possible seemed prudent. But Storz admitted that when they finally decided to pay the kind of salary that top radio people demanded, revenues began to rise because more listeners were tuning in. "We'd rather pay one good man three times what we'd pay for three mediocre ones. Then we get happy people who will give us something extra."

Storz also embraced the concept of limiting the number of commercial announcements in any hour. He knew that more immediate revenue would be derived from the sale of more announcements, but he rightly concluded that such a policy would drive listeners away. Limiting the number of commercial announcements was the better policy in the long run because, when all commercial time availabilities were sold out, it was time the raise the rates!

• THREE •

Forty Favorites in The Big Easy: WTIX, New Orleans

Buoyed by the rapidly growing success of KOWH, in the summer of 1953 Todd Storz completed negotiations to purchase WTIX-AM in New Orleans. It was the second station in his fledgling plan to build a radio station group.

The New Orleans *Times-Picayune* newspaper reported on August 7, 1953, that the Mid-Continent Broadcasting Company of Omaha had purchased WTIX from the Royal Broadcasting Company, whose officers included Brig. Gen. Raymond Hufft, U.S. Representative F. Edward Hebert, and others. "The sale price was not disclosed," said the *Times-Picayune*. It later became known that Storz had negotiated the sale with three checks in his suit coat pocket. One was for $25,000 (about $210,617 in 2012), and two other checks were in greater amounts, if needed to close the deal. The owners took Storz's first $25,000 offer.

WTIX was described in FCC parlance as a Class IV AM station — one whose transmitted power in those days was limited to 250 watts, day and night. That low wattage, and its 1450 frequency high on the AM band, meant WTIX's coverage was limited to the New Orleans metropolitan area. Radio broadcasters referred to Class IV stations as "peanut whistles," because of their limited power and coverage. But unlike KOWH, WTIX was licensed to be on the air 24 hours a day, and the price was a steal, giving Storz-style radio a presence in New Orleans.

WTIX's studios and offices were crammed into little more than 1,200 square feet at 624 Canal Street. Outwardly, it appeared that the minuscule station headquarters matched the transmitter's meager power output. But Storz had big plans for WTIX. He named Bud Armstrong as general manager. Armstrong was thoroughly familiar with KOWH music policy, station stunts, promotions, talent search and selection, and he had become an effective KOWH sales representative. He was the right choice for the job.

Many of the programming practices that had made KOWH a top ranked independent station were installed at "The New WTIX": the top 10 with popular standards, WTIX "News Live at :55," listen-to-win contests, stunts, games, and personalities. At first, Don Loughnane's voice-tracks for WTIX were recorded at KOWH on the production studio disc-lathe and were shipped to New Orleans — along with Johnny Pearson's morning show gags, including his alter ego, "Amanda." Loughnane soon arrived in person to be program director and a mid-day host. He was followed by 1954 high school graduate and KOWH inductee William L. Armstrong (not related to general manager Bud Armstrong). At WTIX, Bill Armstrong became the world's first teenage Top 40 disc jockey. In spite of that, decades later he would be elected to multiple terms as a U.S. senator from Colorado.

Bill Armstrong had been at KOWH for only two months — earning a salary of $300 per month (about $2,527 in 2012) — before being sent to WTIX. Fred Berthelson had just succeeded Todd Storz's "right-hand man" Bud Armstrong as WTIX's general manager when Bill Armstrong arrived there. Bill shared an apartment in the French Quarter with Don Loughnane. Their apartment was near the WTIX studios, on the second floor of the Central Savings building. Bill Armstrong remembers that in those very early days, WTIX was selling some commercial spots for 25 cents each — not even "a dollar a holler."

WTIX would fairly rapidly become a "KOWH — New Orleans-style" radio station. But at the outset, it was a "pale comparison" to KOWH, said Bill Armstrong in Jim O'Neill's memoirs. And in the beginning, the WTIX debut under Storz ownership was complicated by what engineer Dale Moudy termed "a desperate situation." In a 1965 interview with Richard W. Fatherley, Moudy recalled

> the fellow who owned the land [on which the transmitter and antenna were located] was going to kick us off the property. We were on a month-to-month rental. Todd negotiated with the owner of a Chrysler-Plymouth agency to build an antenna system atop a one-story building. Fortunately, we got the project completed and the telephone lines installed [which carried WTIX programming from the studio to the new "Chrysler-Plymouth" transmitter site] on the eve of a telephone company strike. Todd wanted things completed as well as we could do them.

Left unsaid was that initially, the technical facilities were close to being a "chewing gum and bailing wire" operation.

By 1954 — a year after the WTIX acquisition — Storz began very quietly putting out feelers for the purchase of New Orleans station WWEZ. Negotiations dragged on for several years, and FCC approval of the transfer of ownership was delayed until 1958. That delay prompted WWEZ to offer its own version of the Top 40 format between 1956 and 1958, but it was not in the same league as WTIX, nor WTIX's chief rival, WNOE. According to Bud Connell, until 1958 WTIX was number one in the market, boasting a 30 percent share of the audience, with WNOE close at its heels.

WWEZ operated at the low end of the AM dial — at 660 kHz — and with 10,000 watts of power instead of WTIX's mere 250 watts at 1450 kHz. Technically, WWEZ's signal was the exact opposite of the puny WTIX. WWEZ's transmitter site had four antennas beaming the signal east and west along the Gulf of Mexico mainland. The station could be tuned in all the way from Pensacola, Florida to Galveston, Texas. But the towers were located in a swamp teeming with flying and crawling insects, poisonous water moccasins, ever-hungry alligators, and many other varmints looking for a home. According to Deane Johnson, who was WTIX program director in 1965–1966 — seven years after WTIX finally moved into WWEZ's facilities — a rifle was still stationed by the transmitter building door in case it was needed. As discouraging as the swampy location was for staff morale, technically it was an asset: grounding an AM radio tower in water helped its signal to spread out even farther.

When the transfer to Storz ownership was finally approved, the problem became what to do with WTIX's former 250-watts-at-1450 kHz facility? Rather than try to sell it, Storz gave it to the "Orleans Parish School Board for use as an educational station," reported the *New Orleans States* newspaper on February 4, 1958. That move not only generated goodwill among the public and local officials, but also denied the frequency to copycat commercial broadcasters who, in Bud Armstrong's words, would try to "ape" the WTIX format. The transfer of the low-powered station to the School Board followed the precedent set by Storz and his father when they returned the frequency of KOWH's sister FM station, KOAD-FM, to the FCC with the suggestion that it be donated to the University of Omaha.

For the Orleans Parish School Board, the agreement with Storz had an additional sweetener. It provided that the Mid-Continent Broadcasting Company would "pay the operating expenses for the first year up to $25,000 if the board is not satisfied with the operation," reported the *New Orleans States*. Further, the newspaper said that "the board has a right to sell the station at the end of the year." Neither of those clauses was exercised.

However marginal the original 1450 kHz "peanut-whistle" WTIX was as a technical facility, its most enduring claim to fame was the title which was given to one of its music programs: *The Top 40*. That became the consensus term not just for a daily show that counted down the hits, but for the most popular radio programming format of the 1950s and 1960s. How *The Top 40* came into being has been the subject of considerable speculation, and a lot of "conventional wisdom." The actual sequence of events is worth a detailed examination.

For U.S. radio listeners in the early 1950s, the idea of tuning to a station whose entire broadcast day was devoted to "playing the hits" was purely hypothetical. There were network and local programs that played some version of the popular songs of the day, but the majority of those shows aired only once a week, and lasted for an hour at most.

By far the best known of such programs was the weekly *Your Hit Parade*, which had debuted on the NBC network in 1935. John Dunning's book *Tune In Yesterday* recounts that the show featured "the top fifteen tunes of the week, served up by maestro Lennie Hayton, the Hit Paraders, and the Lucky Strike Orchestra." Over subsequent years, there were many changes in format and personnel. The program had a 24-year run on radio which ended in April 1959, when the show defected to television. Dunning states that "the real heyday of the program was in the early 1940s, when America's chief Saturday night mania was to find out which song was number one." But during most of its long tenure, Dunning notes, "*Your Hit Parade* was the nation's only real authority in popular music ... based primarily on readings of radio requests, sheet-music sales, requests to orchestra leaders around the country, and jukebox tabulations." In an era when radio networks were dominant, this singular show became the arbiter of which songs were the most popular. *Your Hit Parade* had also developed the formula of building to a suspenseful climax, saving the top song of the week for the last performance in the show.

On *Your Hit Parade*, it was the songs that got the focus, as they were presented by the program's orchestra and stable of singers, even though the rankings of the songs were determined in part by jukebox play and by sales of records. Each week, the producers attempted a slightly different arrangement of the hit tunes, striving for variety or novelty in orchestration or vocal styling.

Meanwhile, some radio stations (typically outlets not affiliated with a broadcast network) offered short programs which aired the bestselling recordings. Some of these shows presented the hit records in a reverse countdown (i.e., number 10 down to number 1), hosted by a disc jockey. But for the most part, such programs imitated the format of *Your Hit Parade* by playing only the top ten or so hits, plus a few "extras." Sometimes, not even ten songs were played.

Bud Armstrong told Richard W. Fatherley that he first heard hit records in a countdown format while in St. Louis following his 1946 discharge from the navy. He was listening to KWK's local program *The First Five*. The show featured a KWK announcer introducing the records, rather than a studio orchestra and vocalists performing sheet music arrangements — as was the case with *Your Hit Parade*. According to St. Louis radio historian Frank Absher, "KWK's The First Five ... appears to have begun in 1943.... I know it ran at least as long as 1950 ... weeknights from 6:15 to 7 P.M." Local stations such as KWK were mimicking two elements of the *Your Hit Parade* program: the reverse-order countdown to number one, and the time commitment of one hour or less.

The eventual decision of the Storz stations to embrace an all-hit-records, all-the-time format was not as audacious as it might seem. It was pragmatic. When the ratings services of the early 1950s showed strong listenership to an afternoon hit records program on KOWH, Storz decided to offer that same programming during other parts of the day, and the ratings increased in those day parts as well. Eventually, hit records (plus brief newscasts) became the primary program element at KOWH — and later at every other Storz station. But what Storz had recognized is that listeners preferred not to wait until a certain day or night of the week to hear the hits, as with *Your Hit Parade*. They wanted to hear hit music whenever they tuned in.

It can be argued that the Top 40 format as offered on Storz's stations changed radio listening from being an experience in which the musically inclined listener tuned to certain programs to one in which that listener tuned to a certain station. Listeners who liked to hear hit music simply explored the radio dial until they found a station that offered the hits continuously. That probably seems like a small modification in behavior on the part of the radio listener, but it resulted in a seismic shift in the radio industry beginning in the early to mid–1950s.

Todd Storz's eventual embrace of an all-hit-records music format may have had its genesis in an experience Storz had while he was in the army during the Second World War. In an interview published in *Television Magazine* in 1957, Storz said,

> I remember vividly what used to happen in restaurants here in the states. The customers would throw nickels into the jukebox and come up repeatedly with the same tune. Let's say it was "The Music Goes Round and Round." After they'd all gone home, the waitress would put her own tip money into the jukebox. After eight hours of listening to the same numbers, what number would she select? Something she hadn't heard all day? No — invariably, she'd pick "The Music Goes Round and Round." Why this should be, I don't know. But, I saw waitresses do this time after time.

The passage quoted above could be the source of the fable that Todd Storz "invented" the Top 40 radio format in a tavern. In fact, what Storz described in the 1957 article concerns only one concept in "music rotation"—that is, that the most popular songs should be played more often than less popular ones. It was hardly the invention of an entire programming format. But the same *Television Magazine* article quoted WHB's Bud Armstrong cautioning his disc jockeys about the problems they would face because of repeated playing of the most popular records:

> About the time you don't like a record, mama's just beginning to learn to hum it. About the time you can't stand it, mama's beginning to learn the words. About the time you're ready to shoot yourself if you hear it one more time, it's hitting the top 10.

Armstrong's point was that even habitual listeners did not suffer the overexposure to a song that a station's programming staff did. The air talent needed to continue to portray excitement and affection for the music long after they had "burned out" on it.

In a *Kansas City Star* article published 30 years later, Bud Armstrong was unequivocal in again answering the Top 40 origination question, insisting that "Top 40 was not a blinding revelation in a bar." The article further stated that

> the format of playing popular hits, discerned through research with surveys and record store sales, was refined in Omaha over two years. It was only after the Storz group acquired a second station — WTIX in New Orleans — that the format found a name.... After arriving in New Orleans in 1953 to manage WTIX, he [Armstrong] found another local station [WDSU], a network affiliate, that filled time between the end of soap operas and the start of network news, with music; "The top 20," the station called it. "I simply thought that if 20 was good, 40 was better. We called it the Top 40, and [later] when we bought WHB, we stuck it in."

Armstrong was even more assertive in describing the true origins of the Top 40 idea when he dispelled both the myth and its mythmaker. In a July 1996 interview at the University of Maryland, Armstrong said that

> there's an old myth that's totally false, and I want to dispel it right off the bat. The old story that went around — I don't know who fabricated it — but it was a total fabrication — that somehow, Todd, or Todd and I, went to a bar one night and heard that the jukeboxes were playing more of some records than others, and then this great inspiration was spawned to do the kind of format we were doing. That's totally untrue. We were already doing it [at KOWH]. The jukebox operators reports [of which records were being played most] were brought in more to try to balance what we were getting from the record stores than anything else. Sure, any fool who ever went to a bar knew that some records were played more often that others. There was no great inspiration in that. We knew that going in.

The myth that this was some kind of revelation is not true; never was true. I never heard the story myself until 15 years later when somebody was probably peddling it to make a name for themselves.

The facts are these: the term "Top 40" originated in 1953, following the purchase of WTIX, when Todd Storz and WTIX general manager Bud Armstrong heard New Orleans station WDSU playing "The top 20 on 1280" in between NBC network programs, and proceeded to double the number of hit records played on WTIX to 40. From then on, "The New WTIX" was playing the Top 40 records in reverse order every afternoon, building audience anticipation for its countdown climax of the top 10 in reverse order — the same kind of expectation that had been created by the producers of the network program *Your Hit Parade*. That the biggest hits were played near the beginning of the dinner hour did not seem to hurt WTIX's ratings, perhaps because those tuning out were replaced by others tuning in to hear the top tunes. Each week, up to a half-dozen new releases would replace older WTIX records which had run their course, and were thus retired from airplay within the Top 40 rankings.

Whereas many other radio stations played some hit records some of the time, by 1956 the five Storz stations were playing hit records all of the time. The records were selected by each station's program director; there was not a "national" list from Storz headquarters in Omaha. Storz PDs were expected to stay alert to national trends in popular music, but they were free to play a few records that had only regional or local appeal. In the case of WTIX, the records chosen for airing weren't necessarily the consensus national hit versions of the songs.

Bob Walker is a longtime New Orleans disc jockey who is passionate about the recorded music his city nourished, and the radio stations that aired it. Here are three paragraphs from his online essay "Thru The Years..." that explain why the music on WTIX was a little different from the music on other Storz stations:

> In the '50s when vanilla radio stations around the country were playing dreary and sanitized music like "Tutti Frutti" by Pat Boone (!), WTIX in New Orleans was playing it by Little Richard (it was recorded here at Cosimo's Studio, as were the early hits of Ray Charles, Sam Cooke, Charles Brown, etc.)....
>
> When others played "Earth Angel" by the Crew Cuts, we were playing it by the Penguins. They played "I Hear You Knocking" by Gale Storm, while we played it by New Orleans' own Smiley Lewis. They played "I'm Walkin'" by Ricky Nelson, and we played it by our own Fats Domino! ....
>
> While totally segregated musical groups were being force-fed to listeners on the radio, our legendary New Orleans superstar Frankie Ford was singing "Sea Cruise," backed by R&B great Huey "Piano" Smith and giving birth to a milestone song in rock 'n' roll.[1]

Would-be imitators of Storz programming were puzzled by such local autonomy, perhaps imagining that the phenomenal popularity of Storz music programming could only be accomplished through tight central control. On the contrary, Todd Storz hired management-level people he believed could do the job, and then allowed them to operate on their own until problems occurred. That was as true of music selection as it was of any other aspect of station operation.

When teenage-appeal records began vying for airplay alongside older, established artists, and when dozens of new labels began crowding the record bins formerly controlled by the "major" record companies, the number of weeks a song spent on the charts began to shrink. That, in turn, caused the ranking of record titles to change more often. Local listeners' requests to radio stations, tallies of record store sales, jukebox plays, and pressure from record promotion people also caused a faster turnover of record rankings by trade publications including *Variety*, *Billboard* and *The Cash Box*. The latter record trade magazine proudly claimed, "Only those records best suited for commercial use are reviewed by *The Cash Box*"—as if that were a guarantee of quality. All three publications were closely studied by Storz program directors.

Although 12-inch LP records that revolved at 33⅓ rpm and which could contain a dozen or more songs were becoming commonplace in the mid–1950s, it was the small 7-inch 45-rpm "single" discs (with a large center hole) that were in demand by the teenage audience. They were inexpensive, durable, and easy to carry to parties. They allowed teens to buy only the songs they liked — much as today's music fans buy online mp3 versions of only the songs they like — rather than entire albums. Teens had become the primary market for new hit music — and thus prime targets for advertisers.

What was most often on those 45 rpm records were "songs that dealt specifically with youth issues — break-ups, nagging parents, school problems," in the view of David Simons, author of *Studio Stories*. These songs were like a magnet, especially to young female teens, and drew them into record stores following their broadcast on radio, ABC-TV's *American Bandstand*, local TV versions of "bandstand" shows, disc-jockey "record hops," and word-of-mouth. Said Simons, "For the first time, young people comprised the majority of the record-buying public." Radio stations that offered an all-hit-music format "owned" the teenage audience, especially at night, over weekends, during school holidays, and all summer long.

The Storz stations were among the first to capitalize on this drove of young people who became as attached to their new "transistorized" portable radios as most people today are attached to a cell phone, and who listened avidly in their cars — once they were old enough to get a driver's license.

Storz's Mid-Continent Broadcasting Company, and its affiliated sales representatives, dubbed its share of young listeners "the youth market." Reaching this audience on behalf of youth-oriented advertisers would eventually generate tens of millions of dollars for the Storz stations, even though the company was well aware that national advertisers wanted to see ratings showing substantial adult listenership.

Marc Fisher, in his highly readable book *Something in the Air* (pp. 26–27), says

> Storz always packaged his programming as something for the entire family. He wanted his stations to stay out of politics, out of anything divisive. He just played hits and entertained people, he'd say. Like generations of popular culture executives to come, Storz argued that radio had no obligation to promote better taste. "We follow the trend; we do not try to lead it," he said. "The hit tune is the common meeting ground...."
>
> ... Even within the Storz empire, some doubted the new music. "Many of the old-line guys didn't want to run into all the static we got from playing rock 'n' roll," said Steve Labunski. "And it was hard, because you had to dominate the market for it to work. If you had 60 percent of listeners, you had lots of adults in there. But if you had 15 percent of the audience, those were just kids and you were in trouble." Luckily for Storz's philosophy, "if you played the music often enough in the household, the parents would hear it and get curious and then you'd have them. The key was frequent repetition of the music."
>
> But as much as he maintained a brave front to the public, Storz was shaken by the backlash. At one point, he became so frustrated by advertiser skepticism that he had his stations send doctored airchecks — the tapes that gave potential sponsors a taste of the station's sound — to New York ad agencies, deleting some teen tunes and splicing in Sinatra and other more adult numbers. "All of a sudden, the music was a little sweeter," Labunski said. "It was part of a never-ending struggle to get big-name accounts."

Although many people who "came of age" during the heyday of the Top 40 radio format in the 1950s and 1960s feel that the variety of music those stations played then was a sort of "breakthrough" in embracing all styles, veteran Top 40 programmer Bud Connell points out that many categories of music received radio airplay prior to the advent of the top 40 format. Connell originally joined the Storz organization as a disc jockey at KOWH in 1956, shortly before Storz sold the station to William F. Buckley. As will be shown in later chapters, when Connell became a program director at stations that competed with Storz outlets in New Orleans and Miami, he made it very difficult for the Storz stations to acquire or maintain a number one ranking. Later, he achieved a career high by rejoining the Storz organization and becoming program director of Storz's phenomenally successful KXOK in St. Louis. He has a long and broad understanding of what constitutes hit music on the radio.

The selective year-by-year listing of recording artists and record titles that is reproduced below was developed for this book by Connell, based on research in the *Variety, Metronome*, and *Billboard* music trade magazines. Connell's list demonstrates that musical variety was inherent in the recorded music heard on the radio prior to, during, and after the Top 40 era of the mid–1950s through the mid–1970s:

1935 saw a pair of the earliest crossovers [from the rhythm 'n' blues category to general market popular music sales] when two records by Fats Waller made it to the number one position nationally for several weeks. The Ink Spots crossed over in 1939 and continued to do so for years afterward. Quite a few Rhythm & Blues artists followed in the middle and late '40s and the early '50s.

Country Music began penetrating the world of pop from *1947–1949*. In 1947, Tex Williams' "Smoke, Smoke, Smoke that Cigarette" was a massive country hit, and "Open the Door, Richard," by the Three Flames, was an equally massive R&B hit which crossed into pop. In 1949, country star Jimmy Wakely joined Margaret Whiting with "Slippin' Around," and the "big kahuna," Hank Williams, became the first country music crossover superstar.

1950 ushered in a huge and enduring Christmas novelty song "Rudolph, The Red-Nosed Reindeer," by Gene Autry, more country with Red Foley's "Chattanooga Shoe Shine Boy," teeny-bop music with Teresa Brewer's "Music, Music, Music," and even comedy with "The Thing," by Phil Harris.

The *1951* rhythm 'n' blues hit "Sixty Minute Man" by the Dominos was the first record to mention "rock 'n' roll" in the lyrics, In spite of imagery portraying how the singers could maintain rather stenuous lovemaking throughout the night, a few independent stations dared to play it. [Connell did so during his first year as a radio deejay — in Georgia.]

1952 offered a more mature country sound, unique comedy, and a Christmas novelty with (respectively) "Slow Poke," by Pee Wee King, "It's In the Book," by Johnny Standley, and "I Saw Mommy Kissing Santa Claus," by Jimmy Boyd.

1953 introduced audiences to satirist Stan Freberg with a comedy single.

1954 gave us The Crewcuts (teen appeal) and The Chordettes (pre-teen and early teen) and the first release by Elvis Presley on Sun records, "That's All Right."

Bill Haley & the Comets, The Crewcuts, Johnny Maddox, Tennessee Ernie Ford, Fess Parker, The Platters, Chuck Berry, The Penguins, The Cheers, and Fats Domino all broke into the Pop charts during *1955*.

The eclectic mix of Elvis Presley, Bill Doggett, Frankie Lymon, Little Richard, The Teen Queens, Sanford Clark, Patience & Prudence, The Charms, Ivory Joe Hunter, and The Diamonds all had major airplay in *1956*.

To sum up the decades of the 1950s and 1960s, when Top 40 stations were dominant: Select comedy and light teen appeal records were added in the period of 1950–53. Doo-wah and rock 'n' roll entered in 1954–55. Rockabilly began to make appearances in 1956–57. By 1958, the mix was homogeneous and remained relatively stable until the 1964 "British Invasion" by the Beatles and other English acts. For a few years in the mid 1960s, Top 40 music categories again remained mostly fixed—until the advent of drug culture music began its early stab into pop music in 1968–69. More extreme examples of later drug culture songs, hard rock, and metal were reserved for AOR (album-oriented rock) stations and rarely found their way onto true Top 40 outlets concerned with maintaining mass audience. Likewise, throughout the run of Top 40, pure country and pure rhythm 'n' blues selections were heard exclusively on specialty stations.

Bud Connell's credentials as a highly successful Top 40 program director and popular music expert are unassailable. The notion that a radio station could hold first place in a major market by relying on music that appealed only to teenagers, or to devotees of just one music genre, was simply not true. The music heard on Storz's Top 40 stations was more eclectic—and audiences were more diverse—than the "conventional wisdom" about Top 40 seems able to allow.

By 1956, Todd Storz had also purchased WHB, WDGY, and WQAM, as will be related in succeeding chapters. Those stations were successful, but by the time of their acquisition, they (and KOWH and WTIX) all had spawned local competitors. Since the Storz station wasn't "the only (Top 40) game in town" anymore, continuing to build ratings and increase advertising revenues could only be accomplished by out-programming the competition.

When Storz was vacationing in Hot Springs, Arkansas, in 1956, he telephoned Bud Connell while Bud was on the air at KXLR in Little Rock. Connell remembers:

> When he called, I happened to be reading a news item about him published in a recent *Time* magazine. He had given radio a new life, saved it from death by TV, and was being recognized worldwide. I studied his picture as he spoke to me on the phone. He didn't look much older than I, but his image in *Time* made him appear as a giant in my young eyes. Shortly after that initial call, his program director negotiated a deal with me to move to KOWH in Omaha. He was generous for the time, and I would have taken the job for any salary offered. A chance to work for Todd Storz was a piece of serendipity that had fallen to Earth. The next twelve years would be my own rocket ride into programming and management that would forever alter my life. I would learn how to take chances on the instinct that I could second-guess public taste. I called it programming by creative leaps of faith.
>
> In Omaha, I discovered my ability to maintain dominance of an audience. I was acting by rote, recreating what Todd had created—following the leader. But when Todd sold the station (and the on-air personnel with it) a year later [in November of 1957], I found myself working for a young publisher named William F. Buckley, Jr. He had plenty of charisma but no broadcasting background.

Connell had a decision to make. He had eagerly accepted the offered disc jockey position at KOWH because he highly admired how the pioneering

Storz station was programmed and operated. Connell's Omaha colleagues in those days included vice president and general manager Virgil Sharpe, staff announcer Grahame Richards, and program director Bill Stewart. (Both Richards and Stewart would go on to become national program directors for Storz as the stable of stations grew.) But Connell guessed correctly that under different ownership, KOWH's leadership in program innovation would disappear. So Connell decided to phone Texas-based Top 40 station owner Gordon McLendon, and ask him for a job. McLendon owned stations in the so-called "Texas Triangle"— KLIF in Dallas, KILT in Houston, and KTSA in San Antonio. And he was the son-in-law of the ex-governor of Louisiana, James A. Noe.

> Gordon offered me a generous raise over my KOWH salary, and he gave me a choice. I could go to work as a disc jockey for any of the Texas stations, or become the afternoon drive personality on his father-in-law's WNOE in New Orleans. He said the potential for growth was the greatest in New Orleans, and that they might soon need a new program executive. When he revealed that he had just hired Gary Owens [who would later become one of the stars of NBC's hit TV show *Laugh In*] to be the morning show personality on WNOE, I knew he was serious about making a mark in the Crescent City. The overriding factor in my decision, though, was that Todd Storz's WTIX was dominant in New Orleans (even though in 1957 it was still operating with a mere 250 watts), and this would be my chance to beat Todd in the ratings, potentially later returning to Storz Broadcasting as a conquering hero. There was no contest. I took the job in New Orleans and a few months later was named program director. It was my opportunity to win ratings and make my programming statement.

McLendon's national program director Don Keyes oversaw WNOE programming and promotion for a few weeks after Connell arrived, but James A. Noe, Jr.—who then was WNOE's general manager—observed that Connell had the highest ratings on his father's station, and was within a few points of beating his direct competition on WTIX. Middle management was reshuffled, and Connell was named program director. There was no further input regarding programming from McLendon's home office in Dallas after Connell's appointment.

With Connell at the helm of WNOE programming, WTIX's ratings began to fall almost immediately, and Storz national program director Bill Stewart was called in to shore up the station. A slate of new promotions hit WTIX's airwaves, but after each introduction, Connell "covered" WTIX's promotion within 24 hours with something bigger and better. Apparently WNOE listeners didn't tune to WTIX often enough to discover the one-day lead WTIX had. Moreover, WTIX listeners were being attracted to WNOE by a bigger prize, or a more challenging contest, or both. For example, Storz's

"Lucky House Number" was covered by WNOE's "Mystery House." Instead of announcing a different street address each hour for a growing prize, Connell's "Mystery House" was a single secret house with a much bigger prize than WTIX offered. The WNOE prize would go to the first person who knocked on the right door and asked, "Is this the WNOE ten-sixty Mystery House?" When a winner was determined, a new Mystery House was chosen, and the contest started all over again. The advantage, Connell reasoned, was that the Storz promotion was more static, whereas the WNOE version prompted thousands of people to promote WNOE's call letters and dial position to strangers at no charge.

Storz's "Lucky License Number" contest was covered by WNOE's "Rear Window" promotion. Connell purloined the name from Gordon McLendon's stations and from 1954's Jimmy Stewart hit movie thriller of the same name. A listener had to have a WNOE "Rear Window Sticker" on the back window of their car in order to win — if WNOE called out "your lucky license number." So, without stealing the original Storz promotion's "Lucky License Number" name for the window sticker, it was possible to use the concept and the words "lucky license" without fear of reprisal.

If some might have accused WNOE of "buying" an audience with their contests, Connell could rightly point to strong programming as the real reason for the station's ascendance in the ratings. Connell had hired and assigned Jim MacKrell — known on-air as Jay McKay — as News Director. He became Connell's creative sounding board, and was responsible for news scoops, "busts," and exposes. With MacKrell, Connell created *WNOE Intelligence Reports*, the nation's first high-energy hourly newscast, replete with musical "stingers" and mood-oriented music "beds." It was the predecessor of WFUN Miami's *FUNdamental News* and KXOK St. Louis's *Essential News*.

Other program elements such as updated imagery, personalities, contests, and games were specifically created and aired to bolster WNOE's ratings. Examples include "The Mile-High Weather Eye," "The Weather Girl," a hundred hidden plastic Halloween pumpkins each of which was stuffed with prizes, candy, and cash, and the "Sardining Contest," in which the object was to stuff the greatest number of people into a Volkswagen (the winner crammed 26 into one VW coupe!) Connell's evening deejay, Shad O'Shea, chose a wife from applicants on his show, and eventually married her in New Orleans' famous St. Louis Cathedral in a ceremony which was — of course — broadcast over WNOE.

But Storz's WTIX didn't roll over and play dead. In 1959, Grahame Richards — who had been a fellow disc jockey with Connell at KOWH — replaced Bill Stewart as Storz's national program director. With Richards as WTIX's new "idea guy," the competition with WNOE became "blazing hot" as Connell described it:

> WTIX, under Grahame Richards guidance, rolled out a succession of highly creative promotions. In one, Richards hired an Indian from North Dakota to lie in a pit grave on a major shopping center's parking lot. The ten-foot-deep grave was covered by a clear plastic viewing box. WTIX charged an admission (which went to charity) to view the Indian, Wachikanoka, lying in the grave covered with hundreds of live, genuinely poisonous snakes. It was a sideshow brought to the level of a feature performance, and went on for nearly three weeks.

When word got out that WTIX's promotion was coming, Connell employed an Oklahoma deejay — a full-blooded Indian — and renamed him "Charlie Cherokee." To combat the WTIX "snakes in a grave" promotion and herald the arrival of WNOE's newest air personality, Charlie was dressed in full Indian regalia including a head-to-toe feathered headdress. His first order was to check in at New Orleans' famous Roosevelt Hotel carrying his personal pet: a ten-foot-long boa constrictor. Needless to say, Connell had WNOE news director McKrell tip all the newspapers and TV stations about the strange man trying to check in at the Roosevelt with a giant snake. They all "bit" hard, WNOE received massive free coverage, and the other media had to admit that the story had been a stunt to herald WNOE's new deejay. In retribution, the next day Grahame Richards crashed into Connell's studio while Connell was reading a live commercial, and wrapped a boa constrictor around his neck. "Such were the all-fun days of the late 1950s — the height of the radio battles," said Connell, who offered these kudos to WTIX's Grahame Richards: "WTIX was highly active and creative during Grahame Richards tenure as Storz national program director. He brought in the strongest people, contributed the most creative promotions, and gave me the strongest competition of my life."

Connell remembers that WTIX cycled through several program directors during his tenure at WNOE, and that after a period when WNOE went total audience-dominant in May of 1958, Todd Storz interviewed him for the managership of WDGY in Minneapolis–St. Paul. But at just 23 years of age, he was deemed too young for the job.

> With chagrin and indignation, I advertised my availability in *Broadcasting* magazine and attracted offers from *Esquire* magazine, WINS in New York, and Rounsaville Radio. *Esquire* was new to the radio business. J. Elroy McCaw offered me managing directorship of his entire broadcasting empire including New York's WINS. But Bob Rounsaville wanted a manager to open a new station in Miami against Todd's fabled WQAM. I always wanted to live in South Florida. Again, no contest. I took the offered managership and built the new WFUN, Miami. We were top-rated in less than sixty days.

Having been beaten twice by Connell, Storz eventually would offer him programming control over newly acquired KXOK in St. Louis, in the nation's ninth largest city. Connell would jump at that offer. (Both the WQAM vs. WFUN Miami battle, and the KXOK story, are covered in later chapters of this book.)

A different, later perspective on WTIX has been supplied by Deane Johnson. He remembers back to a Spring day in Oklahoma City in 1965, when Rex Miller — the general manager of Storz's KOMA in that city — summoned his production director Don McGregor and his program director (Johnson) into his office and told the pair that "Bud Armstrong wants to see you in Kansas City right away. Here are your plane tickets." There was no explanation as to why. Storz's national program director Bill Stewart met McGregor and Johnson at the Kansas City airport and drove them to the WHB studios downtown in the Pickwick Hotel. McGregor was sent up to Armstrong's office first while Stewart and Johnson went to the hotel bar. Then it was Johnson's turn to visit "the sweat box" (Armstrong's office).

McGregor had just been told he was the new program director of KOMA, and Johnson was told he was moving to New Orleans and would be program director of WTIX. Neither of the men was asked if he were interested in such a move. They were simply told that this is what was going to happen.

As Johnson recalls it, he departed for New Orleans within 24 hours, leaving it to his wife to prepare for the move, get rid of their house, pack up their three young boys, and get to the Crescent City. Upon his arrival, Johnson was met at the plane by WTIX general manager Fred Berthelson. They then headed for the French Quarter, where Johnson checked into the Monteleone Hotel. Berthelson told Johnson that he wanted him not to go to the station until Skip Wilkerson — the departing program director — left nearly a week later. Johnson found this curious, and he later theorized why. It had to do with all of the station "politics and games," as Johnson saw them. Johnson surmises that Berthelson didn't want the departing PD to offer his version of station operations to Johnson.

By 1965, the New Orleans Top 40 market had changed a lot since those feisty days in 1957 when Bud Connell had arrived on the scene to take on Storz's WTIX. Todd Storz's untimely death early in 1964 (to be described later in this book) had left an inspirational void at the top. The smaller and more remote of the pristinely operated Storz stations had begun to show signs of slowing down, of becoming slack in the face of growing competition, and in Johnson's view, nowhere was this more obvious than in New Orleans. His recollection of the situation upon his arrival in the Big Easy in 1965 was that WNOE programming was rather passive, and WTIX's ratings were respectable.

However, Grahame Richards was on his way out as Storz national program director. WTIX's studios had been moved from downtown to the swampside transmitter site near Chalmette, LA. That was part of a never explained economy drive involving several of the Storz stations — which also included a plan to automate the programming. Program automation was tried for a brief period at WDGY, WQAM and KOMA, but fortunately, Johnson says, it was never implemented at WTIX. (To see the "swampside" WTIX transmitter site, studio equipment, and Deane Johnson running a WTIX "remote" broadcast from a mobile studio, go to *www.pbase.com/deanej/wtix_new_orleans*. Click on each thumbnail photo to see a larger image.)

Deane Johnson recalls that WTIX — until several years earlier — had been well programmed under the leadership of Ron Martin, who was rewarded for his excellent work by being moved to WHB as program director. Johnson says, "I don't remember much about the WTIX air staff I inherited with the exception of the evening jock — but they weren't strong. WTIX had a decent news department, and an excellent news director, Charles Ray. We ended up with a pretty good-sounding station, except for the morning show, which never jelled."

Johnson admits that he was puzzled about how to do station promotion for WTIX, because as an Iowan — who had also spent some time in Oklahoma City — he had little feel for the people and the traditions of the Crescent City. The music preferences of the citizenry were also foreign to him and hard to figure out, but Johnson made friends with a couple of local record people whom he grew to trust, and they helped him make adjustments.

In fact, the record people figured prominently in Johnson's introduction to New Orleans' laid-back business style. Says Johnson,

> The radio community in 1965 when I arrived in New Orleans was like nothing I had ever seen before. In most markets we would view our competitors as a mortal business opponent, usually not someone we maintained social relationships with. Not so in New Orleans in the mid-sixties.
> My first shock came only a couple of days after arriving while I was still staying at the Monteleone Hotel in the French Quarter. The Capitol Records regional promotion person sought a lunch so that we might get acquainted. (Lunch in New Orleans isn't just lunch. I found that it meant stopping at a couple of bars before eating and even making several social calls. It had a way of depleting hours of time.) It was suggested we stop in at WSMB and I could meet Marshall Pierce, the program director. WSMB was number one at the time, as I recall, and this seemed a bit unorthodox to me. As we arrived, the door to the studio was propped open and we simply walked in. Marshall was on the air, and the mic was live. He interrupted what he was talking about and started talking to us — continuing live, of course. I was in a state of disbelief. The Capitol Records fellow explained that I was the new program director of

WTIX. Marshall invited me to sit down at a mic and talk about my future plans for WTIX. It was with this episode I began to learn what a major fixture in New Orleans Storz Broadcasting and WTIX were.

That was only the first of several educational opportunities about the New Orleans radio market I would encounter. The next came when I was informed by WTIX manager Fred Berthelson that I would spend Fridays with the various record people. When I suggested this seemed a rather non-productive use of time, I was informed firmly that it was tradition for WTIX to provide this courtesy and I was expected to continue it. It usually consisted of visiting their offices to hear some of their product, then we headed to the Quarter for relaxation at the bar during the rest of the day and evening — on their expense accounts.

WTIX's image in the New Orleans market in 1965 was significant. An example of its stature was demonstrated to me one evening at the famous Al Hirt club, where Al himself was performing on the revolving bandstand. I was there as a guest of David Oreck, who was the distributor for RCA records in New Orleans. (He was the same David Oreck later of vacuum cleaner fame.) At a high point in the show, Al stopped the performance and announced that the new program director of WTIX was in the audience and asked me to stand up and be introduced. He then went on to tell the audience that WTIX played non-stop daily throughout his home.

Friendship among the New Orleans radio folks continued to amaze me. As the sun set each day, many times we would all gather at the Old Absinthe House on the corner of Bourbon and Bienville streets in the French Quarter. The record people would be present for the purpose of public relations and always ended up taking care of the bar tab for the radio folks. Once in awhile Fred Berthelson would pick up the tab, charged to Storz Broadcasting of course. I remember on one occasion ending up with the staff of WTIX's major Top 40 rival WNOE at dinner. Jimmie Noe, owner of WNOE, was hosting. It made little difference that I represented WNOE's major competitor — it was nighttime and time to socialize in the New Orleans tradition.

Bud Connell finds Deane Johnson's descriptions of fraternizing among New Orleans radio rivals in the mid–1960s to be quite remarkable:

Nothing like this ever happened when I was in New Orleans from November of 1957 through November of 1960. The reason: I brought with me to WNOE a Storz attitude and secrecy, which I learned from Todd and the staff at KOWH. We simply did not associate with competitors, and we were courteous to record people, but we kept them at arm's distance except for occasional lunches and dinners. Loose associations among stations and the record industry were discouraged at stations I programmed or managed; however, we did allow the record companies to entertain us occasionally and provide talent for our listener appreciation shows and special events.

It eventually became apparent to Johnson that WTIX's "internal politics" were the biggest challenge to the station's success. For example, he doesn't recall there ever being a sales meeting held while he was there, and there were no management discussions about getting new business, doing promotion, or winning ratings battles. Storz national program director Bill Stewart came to town a couple of times, but he never ventured out to the station, choosing to "hold court" at the Playboy Club instead. "It was the 180-degree opposite of KOMA under Jack Sampson," Johnson said, referring to his previous post. "I admit to being distracted by all of this and not doing a good job at WTIX, but we did get the station back to sounding pretty good with a new talent lineup. It actually ended up sounding like a Storz station once again." He adds that the saving grace during that mid–1960s era was that WNOE— WTIX's cross town Top 40 rival — was not strong at the time and as a result, WTIX maintained passable ratings.

For Deane Johnson, Hurricane Betsy was the crowning blow. In September of 1965, the eye of the huge storm passed over the radio station and the Johnson's house with 145 mph winds. Three of the four towers in WTIX's directional array fell into the swamp, and two of them were never found. For a week, the only way to get to the station was by boat. From the day Betsy hit the city until the day Johnson left the station almost a year later in the early summer of 1966, WTIX operated at significantly reduced power supplied by a diesel generator sitting in the parking lot. Several years later, the studios and offices were finally moved out of the swamp and back to downtown New Orleans.

It is clear that during its heyday in the 1950s, WTIX had earned the loyalty of popular music radio listeners, and the respect of other radio stations. If for no other reason than that the Top 40 Format was invented in the Big Easy, WTIX was a "big" deal. But keeping it on top hadn't been "easy."

• FOUR •

Building the Flagship: WHB, Kansas City

WHB in Kansas City was truly one of the nation's "pioneer" radio stations. That it debuted in the very early days of U.S. radio broadcasting is evidenced by its three-letter station identification. The U.S. Department of Commerce had authority over broadcasting in the medium's earliest days— the Federal Radio Commission would not be formed until 1927, and its successor, the FCC, would not exist until 1934. In the early 1920s, the Commerce Department incorrectly assumed that the demand for broadcast stations would not exhaust the possible pairs of alphabet letters preceded by a W or K. Call signs beginning with the letter W were generally supposed to go to stations east of the Mississippi River, and K signs to the west, but WHB (and a few others) remain exceptional even in that small matter.

WHB began broadcasting on 833 kHz in April of 1922, and received its license on May 10. The Sweeney Automotive School was the original licensee. As William James Ryan points out in his article "African-Americans in Local Broadcasting: Kansas City, 1922–1982":

> Station ownership is important for at least two reasons relevant to this history. In those early days, radio was a means of promoting the goals of the licensee. There was no advertising as we know it today. Rather, a station was identified with a particular business or institution which, in turn, financed the entire operation. Programs were designed to serve the needs of the owner. Most radio talent received no pay but were glad to get the free publicity on this exciting new mass medium.

Ryan says that August of 1922 marked the probable first broadcast of African-American artists on a Kansas City radio station. He asserts that from its first year, WHB programming "had the most diverse cosmopolitan list of artists of any station in the city." The talent roster included Tutt's Colored Male Quartette and Jazz Orchestra, as well as Hawaiian music, the first women's show in Kansas City, and religious services including Jewish talent. Music from the jazz nightclubs in downtown Kansas City was relayed by "remote" telephone lines to WHB's studios just a few blocks away. During the early years, WHB transmitted on 730 kHz and then 850 kHz, finally settling on 710 kHz in 1946.[1]

The Cook Paint and Varnish Company bought the station in 1930 and operated it until the sale to Storz in 1954. In the early 1930s and 1940s, WHB continued as the home of diverse local radio in Kansas City, and Twelfth Street—which ran from Main Street in the downtown area east to Brooklyn Avenue—was rife with jazz joints, speakeasies, and hideaways.[2] The Scarritt Building, a "Chicago School skyscraper" which housed WHB's studios, was only three blocks away at 819 Walnut Street. The interior boasted mahogany wall paneling, marble wainscoting, and mosaic tile floors. The hallway and studios were lined with pictures of celebrities, Kansas City notables, and movie stars who had visited WHB. Count Basie led a studio band for WHB— and his was just one of two bands the station maintained. WHB's studios and offices occupied the top two floors of the Scarritt, which for all of its opulence had never been air conditioned. "Hell, we've never needed it!" the oldtimers insisted, though the station was brutally hot in the summer. But in the early 1950s, the staff was tired, and ready to retire. They had weathered the depression in the '30s, war in the '40s, and finally, in the early '50s, they were facing the arrival of television.

Most of the pictures and memorabilia from WHB's heyday were still on display when Jack Sampson arrived there in 1950 as a brand new account executive. "Hence," he recalls, "we young ones agreed that the call letters stood for 'We Have Been.'" Sampson was a recent graduate of Kansas State Agricultural College—now Kansas State University. His experience in radio prior to WHB consisted of just four months in sales at KAYS in Hays, Kansas. He started selling advertising at WHB for a 1950 salary of $60 a week (about $464 in 2012), with a promise of a raise if his level of sales was "reasonable." Sampson remembers that his first sales commission check from WHB was for 17 cents.[3]

Both the Cook Paint and Varnish Company which owned WHB, and the Midland Broadcasting Company which owned KMBC radio, became embroiled in a fierce competition for channel 9—the last remaining VHF television frequency in Kansas City. Because the FCC was unable to determine the better qualified applicant, the two companies were forced into an unusual shared operation agreement. Beginning in August of 1953, WHB and the

Four • Building the Flagship   57

owners of KMBC radio (980 AM, now KMBZ), began shared operation of a single transmitter on channel 9, alternating 90 minutes of airtime originating from the separate studios of KMBC-TV or WHB-TV. Less than a year later, Cook Paint and Varnish bought out Midland Broadcasting Company, ending the shared-time agreement. The remaining television station would be known as KMBC-TV. Under FCC rules, WHB radio would have to be sold off. Most of the staff wondered if they were to be transferred to TV, or stay with TV. Jack Sampson was by then selling ads on both media. He was told he would remain with WHB, along with one other salesman. All of the other sales people were to sell for TV. Naturally, Sampson felt disappointed — not knowing then it may have been the best break of his life.

In March of 1954, Don Davis — the long-time president of WHB, along with general manager John Schilling — called the staff together in the main studio. They announced that WHB had been sold, and introduced Todd Storz and Bud Armstrong of Mid-Continent Broadcasting Company as the purchasers. At that time Storz was all of 29, and Armstrong a mere 26 years old. Immediately after the meeting, Sampson approached Armstrong to apply for the job of local sales manager. Armstrong's answer was direct: "Not now!"

Storz had paid $400,000 (about $3,334,818 in 2012) for WHB, a price five times greater than he paid for KOWH in 1949, and sixteen times greater than the price for the original low-powered WTIX in 1953. However, WHB was a fulltime signal at 710 kHz — the preferable "low-end" of the dial, with coverage that extended north to Omaha, Nebraska, south to Joplin, Missouri, east to Columbia, Missouri, and west to Salina, Kansas. But on June 14, 1954, when Storz's Mid-Continent Broadcasting Company took ownership, WHB's ratings were so bad that the jokes went "We're number 7 in a 5 station market!" or "WHB has less audience than taxi radio." WHB programming was a hodgepodge: there were hillbilly bands, a cowboy singer, the Mutual network for news, soap operas, programs designed for housewives by WHB's women's director, play-by-play of the Kansas City Blues (with Mickey Mantle), coverage of University of Kansas and University of Missouri football and basketball, a daily church program, a daily remote broadcast from Kansas City's Country Club Plaza, a "Thought for the Day" feature, and announcers who referred to fast tunes as "lilting bouncers." The program director was reputed to be a drunk. The news department was tiny, but newsman Charles Gray elected to stay at WHB, and his wife, Barbara Hanna, became the continuity writer. Everyone else shifted over to TV. Not one of WHB's original announcers (self-styled DJs) remained.

At his first meeting with the WHB staff, Storz announced that Bud Armstrong would be the new general manager. (Deane Johnson recalls in-house scuttlebutt that Armstrong had been promised a lifetime job if he made WHB a success.) Storz and Armstrong outlined other changes: Affiliation with the Mutual network would be cancelled, as would all of the sports, and the station would be moving to the Pickwick Hotel — with new air-conditioned studios. Sampson recalls that when that last item was announced "a cheer went up! They said WHB was going to have an all-music format like KOWH in Omaha, although nobody at WHB knew anything about Omaha, and even less about WTIX in New Orleans which Storz had recently acquired." Sampson admits his initial impression of Todd Storz was skeptical. "Omaha was that little town up the river. What could Storz know?" But in those early days as the new owner of WHB, Storz earned Sampson's admiration by being "pretty much hands-on all the time."

George W. "Bud" Armstrong at WHB. Bud Armstrong was Todd's lifelong friend, and Todd trusted him to manage the installation of the Storz Top 40 format at each additional station they acquired. This 1956 photograph of Armstrong was taken in Kansas City when Armstrong was visiting WHB.

Hiring of air personnel started in the summer of 1954, with Johnny Pearson coming down from KOWH to be WHB's program director. Peter Tripp was a WHB disc jockey who later would be on the air in New York City. Jack Sampson remembers that they wanted Sandy Jackson to join the WHB air staff because he was attracting huge audiences to KOWH in Omaha, but the move wasn't approved by "the home office."

The Storz Top 40 format went on WHB's air in July of 1954. The countdown show — hosted by Ron Martin — was a three-hour program featuring the top 30 in reverse order, taken from WHB's "40 Star Super Hit Survey." The top 10 countdown was the third hour lead-in to

Four • Building the Flagship

teen-appeal programming which followed in the evening. The other Storz stations would use a similar afternoon countdown format to adhere to a principle that had been proved successful at KOWH: Bring the audience to a top 10 conclusion.

Storz and engineer Dale Moudy were in the station much of the time during late 1954. Sampson recalled that "under Cook Paint and Varnish, [the station's equipment had been there for 20 or 30 years. So a big part of the transition is that we got all-new equipment." Moudy was also installing and testing his own personally developed devices to give the station a distinctive Storz sound. But WHB was the first station where Storz had to deal with labor unions. For instance, Moudy, who was not a union engineer, had to obtain a special permit from the union to "work with the tools" so he could help in the installation of new equipment. And when he had finished installing the echo and reverb units, the local unions brought up another problem: Would the announcers (who were affiliated with AFTRA — the American Federation of Television and Radio Artists) push the button to activate them, or was it the jurisdiction of the union engineers affiliated with IBEW, the International Brotherhood of Electrical Workers? Sampson doesn't remember how Armstrong finessed the engineers, but the dispute ended with the announcers controlling the echo and reverb buttons — which was the only way for the timing and duration of the effects to be accurate.

From an interview conducted by Richard W. Fatherley, here is former WHB newsman Charles Gray's take on installing the Top 40 on WHB:

> It was done by Bud Armstrong. He was the man who brought the plan in and executed it. Todd Storz was in and out, but Armstrong was the man who was on the scene. He was in charge of "The Army of Occupation." The Storz company brought in new equipment, new ideas, *better* equipment. The engineers appreciated that. But they had a little culture shock at first, because the rule was, when you worked at a Storz station — even as an engineer — you worked in a coat and tie. [To see photos of WHB's studios and on-air staff in the 1950s, go to *www.pbase.com/deanej/whb__kansas_city*.]

By the winter of 1954–1955, WHB's original turntables had been converted so that they could play the new 45 rpm discs (with their large center hole), instead of the long standard 78 rpm records. New WHB jingles and musical news intros were introduced, and the new air talent began to sound more modern and have more fun on the air. Charles Gray still comprised the entire news staff. Storz had inherited expensive contracts with the Associated Press, United Press and INS news wire services. To make that triple outlay of cash into a virtue, the station produced a promotional announcement that proclaimed, "WHB has 15,000 newsmen around the world around the clock to serve you!" One day a listener called in and asked to speak to the newsman. "He's out to lunch," replied the receptionist. The response was, "All 15,000 of them?"

In 1954, WHB's initial promotional contest was "Lucky House Number," in which the DJ would pick a street name, then derive the house number at random. The person who lived at that number had one hour to call in and claim the jackpot, which went up $10 every time the call was made. Sampson remembers that he had an undertaker advertiser who wanted WHB to sponsor a "Lucky Hearse Number" contest, but they just couldn't bring themselves to do that. "Imagine 'Win your funeral!'" he laughed.

WHB's first big blockbuster promotion came in October of 1955 — another Treasure Hunt. This time the finders would be keepers of $2,000 (about $16,786 in 2012). Newsman Charles Gray remembers what happened:

> No one knew exactly what to expect. It took the town by surprise. They did it on a Sunday. It began in North Kansas City, then shifted to Kansas City, Kansas, back to Missouri, and ended up in Loose Park. A turtle was there with a number on its back in gold paint. Several thousand dollars were in the hopper for the winner that day. It did create a great deal of traffic congestion. It brought a lot of attention to the station. The station already had a lot of attention anyway.

WHB studio supervisor Roy Nonemaker, who was travelling with Armstrong and Storz in WHB's "mobile news unit" (a station wagon painted with the station's call letters and dial position), also recalled the event:

> Traffic was tremendous. The cops came down there. I heard the sirens. Todd said, "Stop the station wagon here!" He got out and left us. He thought they were going to arrest him and put him in jail, I guess. When the thing was over, Armstrong said, "Roy, I know one thing. People like money!"

The WHB Treasure Hunt stunt put an end to any assertion that WHB was merely a "kids'" station. Thousands of cars driven by thousands of adults proved beyond any question that WHB was everybody's radio station, young and old alike.

Storz spent a great deal of time with Armstrong at WHB, but both frequently talked with the staff, too. Sampson remembers that Storz seemed to love the pace of the station, and his joy and enthusiasm were infectious. During the 1955 "Treasure Hunt" promotion, he was with the staff all of the time. Sampson says it was a pleasure to go to work every morning. Even star disc jockey Johnny Pearson, who could be very moody at times, loosened up and had fun.

Marc Fisher in his book *Something in the Air* paints a slightly different picture of Storz around the time of the WHB acquisition:

Four • Building the Flagship

He was obsessed with radio and everything new, collecting every gadget that came on the market. He remained a ham operator, checking in with other early adopters across the land. He traveled the country to listen to unknown stations in hopes of gleaning a new idea or discovering a bright new personality.... A thin, energetic man with piercing brown eyes, close-cropped brown hair, and a luminous smile, Storz came off as something of a radio geek, but those who worked closely with him saw another side: His frequent travels to visit his stations were also a quest for extracurricular social engagements [p. 14].

Fisher went on to quote longtime Storz manager Steve Labunski on Storz's programming focus: "He cared very little about sales, never went to sales meetings. He cared about programming. He listened constantly to tapes of his stations and others. He'd come to town a day early and not tell you, so he could listen and make sure you were doing everything right" (p. 15).

In less than a year under Storz ownership, WHB's share of Kansas City radio listeners shot up to almost half the available audience — and stayed there. As a result, it became a "hot" buy among city ad agencies. Sampson characterizes WHB's progress under Storz ownership this way: "We took off like a jet airplane. When ratings came out in the fall and winter of that year (late 1954–early 1955), we went up very fast. It was just exciting. We used to say we sold more ads accidentally than we had on purpose before." But the sales staff had to scramble to learn how to interpret and use the ratings books. Hooper and Pulse were the main ratings sources, and Conlan, an Iowa company, also rated Kansas City listening. All of them reported WHB's audience dominance. The station's annual revenues exceeded $2 million (about $17,150,373 in 2012), allowing Storz to retire its purchase cost in just fourteen months. WHB business manager Ray Lollar recalled that Storz once exclaimed, "There ought to be a law against making this much money!"

Jack Sampson admits that he had not been a successful salesman in the pre–Storz years. "I had been pleased, before Storz took over, to sell $4,000 worth of advertising in a single month — which didn't happen very often. Now I was billing $10,000, even $12,000 per month. In 1956, I sold over $150,000 in one year." (That would be about $1,240,472 in 2012.)

Armstrong taught Sampson that increasing advertising rates was a very good thing. The fact was, every time WHB raised rates, the station sold more time, not less. The other stations in Kansas City tried to belittle WHB, saying "rock 'n' roll won't sell," or "WHB has nothing but teenagers," or "crew cuts are the symbol of teenage decadence." But Sampson remembers that "It seemed the more they battled us, the better we got." Asked years later by Tom McCourt if he encountered problems selling time on the new WHB because of its image of appealing mostly to teenagers, Sampson replied:

Sure. People would say that Elvis Presley was a bad influence on the youth, and that Bill Haley and the Comets were wild and crazy. So, sure, there were advertisers that wouldn't buy WHB because they thought the music was all teenage and wild — they just didn't like it. And that's what the other stations tried to use against us, as a competitive sales effort. But when we started getting the ratings — Hooper didn't give us demographics, but Pulse did, and later [so did] Nielsen — they showed that WHB actually had a huge audience up to age 49. So when we marketed the station, we were very successful in proving we had a young-adult audience. By the summer or fall of 1955, we were pretty much handling everybody. We had success stories by then. A couple of the ad agencies were very active with us and used us constantly and successfully. [Station] promotions were good. And the news was good — even though it was only five minutes per hour, it was pertinent and up-to-the-minute and well-delivered.

When asked if any particular category of advertiser was more amenable to buying time on WHB, Sampson replied that automobile dealers were a big source of income, along with movies, soft drinks, and beer. But housing developments were big too, because the G.I. Bill had encouraged the construction of new subdivisions all over the Kansas City area.

Sampson remembered there being a competitive environment within the station. "It was fun, and we all liked each other, but we were very competitive. In the summer of 1956, I got to be [local] sales manager, and we had five or six people on the sales staff — some good kids, who have done very well in broadcasting." As station manager, Armstrong handled all of the national sales.

Armstrong was also responsible for turning an off-the-cuff remark into an enduring slogan for WHB. One day, he overheard one end of a telephone conversation between WHB sales representative Jim Miller and a client who had asked what the letters "WHB" stood for. In fact, the three ancient letters had no intrinsic meaning, but Miller replied, "Why, didn't you know? We're the '*W*orld's *H*appiest *B*roadcasters.'" Armstrong began waving his hands and saying, "That's it!" He immediately directed that the phrase "world's happiest broadcasters" be scheduled every half hour, around the clock. But it is likely that one of the reasons this episode has been remembered is because Armstrong was not always so amenable to new ideas.

Jack Sampson is emphatic in his assessment of Armstrong:

Bud Armstrong was a great general manager. He was very smart and quick, and knew the Storz formula backwards and forwards. Todd insisted all of his GMs know the programming. Bud kept us on our toes, and seemed to know exactly what was going on at all times. One morning the owner of the Pickwick Hotel saw Bud arriving at work at about 10 A.M., his usual time, and said, "Bud, here you are getting in so late — why, President Eisenhower was at work in the White House at 7 A.M." Bud's reply was, "Yes, but he doesn't have the staff I do." Take

Four • Building the Flagship

that John Foster Dulles! Bud loved to hold forth in the Pickwick Lounge after work, and I think I learned more about Storz and radio in those sessions than at any other time.

Sampson's "upper echelon" understanding of Storz operating procedures would allow him to become a Storz general manager in 1958, and a vice president of the company in 1961.

Richard W. Fatherley was a strong admirer of WHB. He always referred to it as "the flagship" of the Storz stations, knowing full well that KXOK in St. Louis — which would be Storz's final station acquisition — had more listeners and made more money than WHB. By using the term "flagship," Fatherley did not mean that WHB was the radio equivalent of the vessel that carried the commander of a fleet, but rather that in the 1950s WHB became the best and most important example of what a Storz station could be. KOWH was a daytime-only station in a city of 366,395 (1950 census, Omaha/Council Bluffs) that managed to attract huge audiences by concentrating on playing popular music recordings. WTIX in New Orleans (population 579,445, 1950 census) had refined the music format by concentrating on the forty most popular records of the day, and succeeded in garnering high ratings despite its meager power output of just 250 watts during its first four years of operation. WHB in Kansas City (population 456,620, 1950 census) was not hampered by either of those technical deficiencies. It was the synthesis of everything the Storz organization had learned about how to operate a completely successful AM radio station.

In terms of signal coverage, WHB was an excellent facility. Its low end of the dial position at 710 kHz and its non-directional antenna system allowed the station to be heard clearly as far north as Des Moines, Iowa; as far south as Joplin, Missouri; as far east as Columbia, Missouri; and as far west as Salina, Kansas. If it were overlaid on a map, WHB's signal "footprint" would have been almost as big as the entire state of Wisconsin. There were claims that WHB could be heard in six states during the daytime hours, and ten states at night.

So far as programming was concerned, WHB did not have to go through an extended period of "experimentation" as KOWH had, nor did the station alter its music playlist to accommodate local tastes, as happened at WTIX. When WHB went on the air as a Storz station, the programming and promotion that wowed listeners in Missouri, Kansas, Iowa, and a small part of southeast Nebraska had already been proven to work.

There is another possible reason why WHB came to be considered "the flagship" of Storz's stations: the advent of rock 'n' roll music coincided with the debut of WHB's revamped format. However, the new music that appealed so strongly to teenagers did not start out with that name. It is likely that Bill

Haley and his Comets contributed the word "rock" to the moniker for teen-appeal music with his song titled "Rock Around the Clock." It was recorded in April 1954, but did not receive much airplay then. In about July of that year, Haley and his band recorded "Shake, Rattle and Roll," which may have been the source for the second word in the phrase "rock 'n' roll." But it wasn't until Haley recorded "Dim, Dim the Lights" in November 1954 that his band began to get significant radio airplay, especially from Alan Freed on his nighttime program from Cleveland, Ohio. According to Edward Dixson's customer review of Haley's "Rock Around the Clock" album on Amazon.com's website, "Dim, Dim the Lights" became "the first 'rhythm 'n' blues' song recorded by a white artist to not only get on the R&B charts but land in its Top Ten." As a result, disc jockeys began to play Haley's earlier releases: "Shake, Rattle and Roll" and "Rock Around the Clock."

It was unlikely that the new WHB went on the air with the intent of becoming a rock 'n' roll station. But if WHB intended to maintain the principle of playing the top musical hits as KOWH and WTIX were doing so successfully, there was no choice but to play the hit rock 'n' roll records.

In 1956, a person from the McLendon Stations—possibly Gordon McLendon himself, but more likely one of his program directors—visited Kansas City to listen to WHB and to talk with Bud Armstrong. A memorandum of that meeting titled "A Confidential Report on WHB" was discovered by Dave MacFarland in the extensive McLendon Policy Books archive in Dallas. The research was not clandestine, and the report reads more like a transcript of a conversation. But it does contain insights into Storz operations and Armstrong's thinking at the time. The most relevant portions of that document are as follows:

> [p. 1, 1st paragraph] WHB presently has a maximum of 15 spots per hour. This includes only one-minute spots plus two quickie 20-second breaks on the half-hour. In other words, 13 one-minute spots per hour, plus two 20 second quickies on the half hour. Armstrong got a little worried about the possible effect it would have on his audience, and the fact that KCMO was making some noises about bringing in some outside disc jockies, so he decided to cut back to 15 spots per hour.
>
> [p. 1, 4th paragraph] All accounting is done in Kansas City for WHB and likewise for all Storz stations in each station's home city. The only contact between WHB and the home offices in Omaha is a monthly profit and loss statement with is sent to Omaha for purposes of overall corporate accounting. There are no weekly or daily reports. Both Storz and Armstrong can write checks on every one of the four bank accounts [KOWH, WTIX, WDGY and WHB].
>
> [p. 2, 5th paragraph] WHB is nicely carpeted around the executive offices, but not back in the studio portion. This station is beautifully equipped. As already

noted in other memos, they have practically no record library. The record library consists of the Top 40 and a few old standards, and that's it. Johnny Pearson, the morning deejay, is the program director.

[p. 2, 6th paragraph] They have some new, short, singing station identification breaks which the musical groups have recorded for them free. They have only one newsman. Armstrong has union problems which make him hesitant to go all out in news until he has to. He says he has a hunch that he will have to someday in spite of the union problems. They of course have union engineering and announcing departments. Armstrong says that the first station doing the music and news formula right can maintain its dominance in the market permanently if it is the equal facility to any other station. He says that he and Storz don't want to go into any markets that they can't dominate, and dominate quickly. He says that this wouldn't be possible in a market like Los Angeles where he says the rate structure is surprisingly low, and will prevent any station from having the money to achieve more than about eight percent to ten percent of the audience...

In the case of Cleveland, when you start buying the top two men in the market, you're talking about $100,000 a year [equivalent to about $836,996 in 2012]. He says that St. Louis would be likewise as far as buying off the top men. He says it's a far better market ratewise. He says that the stations have been able to keep their rates up pretty well there, and that the big trouble in St. Louis is that the union has been allowed to run away with things. He agrees completely on WGMS and the Washington [D.C.] market in general. He says it comes as close to being without competition as any market in the country, and that Chicago is next...

[p. 3, paragraph 4] He says that so far, KOWH has lost only a couple of points in Hooper share-of-audience to now-independent fulltimer KOIL. KOIL is second with about 16 percent of the audience, but will now get a little more commercial business and will begin to level off. He says that this is what has happened to us (WNOE) in New Orleans; i.e., that we went up sharply and then got some commercial business and leveled off. We [WNOE, which was being programmed by McLendon for Gordon McLendon's father-in-law, James A. Noe, who was the owner] were going up sharply, in his opinion, just as long as we didn't have so much commercial and WTIX was overloaded. But when both stations got to about equal and comparable business, we leveled off. I do not agree that this is the reason for the leveling-off, as I think it is a deficiency on the part of the disc jockies at WNOE...

[p. 4, 2nd paragraph from the bottom] Armstrong follows the same procedure we do of removing the disc jockies from competitive stations by simply sending them to another Storz station in another town.

[p. 4, one paragraph from the bottom] The Storz group is shooting for seven stations.

[p. 4, last paragraph] Kansas City has always had notoriously low tune-in, but there are some encouraging aspects also. The daytime tune-in in 1948 in Kansas City, which was before television, was 18 percent on a Hooper. It is still around 14 ½ percent and slowly climbing. The nighttime about a year ago, the last published figures show, was between eight percent and nine percent. Armstrong said that Hooper is his bible, and he believes it completely and implicitly, and that other services are useful to him only for sales purposes if he shows well, and he does. His present share of audience in Kansas City is 49 percent both morning and afternoon.

[p. 5, a middle paragraph] He is still using a jukebox for operation between 12 and 6 [A.M.], but no talk except transcribed announcements.

[p. 5, another middle paragraph] Storz buys stations on the formula of top facility, plus good rate market, plus Top 40 and Lucky House Number, plus no competition. They would have to have no competition in Kansas City because it is one of the dullest sounding stations I have ever heard. It just has the music and news formula in its purest form, and that is about all.

[Last paragraph] Armstrong does not believe in doing anything new at WHB or pulling any tricks out of the bag program-wise until some other station shows in the Hooper that it is beginning to be competitive, and at that point he always pulls something new out to do. Otherwise he rides the status quo.

Deane Johnson disagrees with the comment in "p. 5, another middle paragraph" above, that WHB was "one of the dullest sounding stations I have ever heard. It just has the music and news formula in its purest form, and that is about all." Says Johnson:

I don't think they held back much on the music; I think the "class" came from the mature personalities and their toned-down presentation. As I remember, they didn't do all of the stunts and loud stuff. There was no shouting, mostly just personalities, music, and lots of commercials. WHB didn't have a lot of competition in the Top 40 arena — mostly just KUDL as I remember, and that was a weak facility technically. WHB didn't need to make noise to get noticed. They had staked out their position before anyone in the market realized what was happening and just held onto it as the originator. WHB didn't do anything special other than sound darn good.[4]

Johnson also commented about why WHB was so "laid-back" and "classy." He recalls Armstrong saying, "I see no need to whip my horse when he's in the lead."

Storz had been buying full-page advertisements in radio broadcasting trade magazines such as *Broadcasting-Telecasting* and *Sponsor* since the days of the meteoric rise in listenership to KOWH. He occasionally bought space on the front cover of both magazines — signaling that the Mid-Continent Broadcasting Company was well-established and had deep pockets. The readers Storz was seeking were "time-buyers"— the decision makers at advertising agencies who determined where and when to place radio advertising for their clients' products. The display ads for "the new WHB" pointed to the success advertisers had enjoyed on KOWH and WTIX, with the theme of "getting in on the ground floor" as WHB built its audience. And advertisers did line up for a turn on WHB's air. Here are partial descriptions of some of the trade magazine advertisements for WHB in the period November 1954 through December 1955:

Four • Building the Flagship

November 1, 1954, *Broadcasting-Telecasting*: Big Switch! At midnight, October 17th, WHB completed the switch from network to independent operation. Now WHB has 24 full hours a day to transmit the kind of radio which has already started the big switch in Kansas City listening....

November 29, 1954, *Sponsor*: Big Switch! WHB switches to independent operation and Kansas City listeners are switching to WHB. Unburdened by a lot of programs only *some* people want to hear, WHB now fills 24 hours a day, 7 days a week with what *most* people want to hear....

December 20, 1954, *Broadcasting-Telecasting*: Double Take-Over! June 1954, Mid-Continent Took Over WHB.... And now ... WHB TAKES OVER KANSAS CITY. It happened in Omaha and it happened in New Orleans! Now Kansas City makes three leaders for Mid-Continent! Hooper says WHB is first in the morning, first in the afternoon, first all day with 35.7 percent of the available audience, twice the next station's share....

January 24, 1955, *Broadcasting-Telecasting*: WHB TAKES OVER KANSAS CITY. Now one station dominates the fabulous Kansas City market. WHB, with 38.1 percent of the audience, has more than twice its nearest competitor...

February 21, 1955, *Sponsor*: [There is a drawing of a steer divided into the typical "cuts of beef" except that these show the shares of other radio stations, labeled A through G. There is a very large area at the steer's midsection, on top of which is a table model radio. Words are printed over the steer's large midsection] This is the way they cut up the radio audience in Kansas City ... and WHB gives you the prime cut with 43.5 percent of the all day audience. [Station "A" had only 17.4 percent, and the other segments went down from there.]

March 21, 1955, *Sponsor*: Nearly half of Kansas City is yours on WHB. 48.9 percent to be exact. The other half is shared by seven other radio stations ... Note that WHB's share of audience is nearly 3 times that of the second station. This did not happen by chance. It is the Mid-Continent formula at work. Listeners have been drawn to WHB by the music and news they like, purveyed by the kind of personalities they like and respond to. Advertisers, too, are responding. In February, WHB served 162 separate advertisers — double the number on hand when Mid-Continent took over just 9 months ago.

April 11, 1955, *Broadcasting-Telecasting*: UNANIMOUS. There's unanimity in Kansas City: No matter how you count the audience, the no. 1 station is WHB. *This* is what Mid-Continent programming, ideas and excitement have achieved for WHB! All three national surveys — PULSE, HOOPER, TRENDEX — give WHB the top daytime spot with ratings as high as 48.9 percent...

Sponsor, October 31, 1955: HOOPER says it. NIELSEN says it. PULSE says it. TRENDEX says it. WHB is running away with Kansas City's radio day. Have a pet rating? Doesn't matter. A.M. or afternoon? Doesn't matter. WHB is first in every time segment per every rating service....

Broadcasting-Telecasting, March 26, 1956: IT'S WHB'S REGION, TOO. 263 1st place ¼ hours out of 288 ... 25 second place ¼ hours ... and nothing lower! That's what Kansas City Area Pulse says about WHB for 594,700 radio homes in 66 counties of 3 states....

Broadcasting-Telecasting, November 12, 1956: [A counter-ad from the Meredith Radio and Television stations showed a cartoon of a guy with long sideburns, dressed in a loud striped suit, strumming a guitar, and singing into a microphone. Alongside him, there was this caption] In Kansas City, if you want to sell the rock-n-rollers, there's a place to go ... [The bottom panel was a drawing of a contented family in a living room, listening to the radio, with a second caption:] "but if you want to sell the whole family, it's KCMO radio. [Apparently, KCMO didn't have any audience numbers that they wanted to feature.]

Sponsor, January 12, 1957: [There is a photo of two businessmen in the back of a taxi. The caption next to one says:] "Ratings make them sign the first time." [The caption next to the other man says:] "But it takes results to make them renew." [Under the photo, a headline says:] At WHB ... 87 percent renewal. [The beginning of a text block reads:] "87 percent of WHB's largest billing local accounts in 1955 have renewed in 1956 ... it takes *results* to make local advertisers come back for more. And WHB is Kansas City's *results* station. So much so, that WHB has a higher percentage of renewals for both local and national advertisers than any other Kansas City radio station.

Sponsor, March 23, 1957: [There is a photo of a woman standing at her front door as a neighbor begins to cross the street with children. There is a radio on a table near the front door. The caption reads:] The kids have left for school.... *Now* what station will she listen to?" [The text under the photo continues:] All-new surveys show again: WHEN THE YOUNGSTERS ARE AWAY, KANSAS CITY RADIOS STAY ... WITH WHB. Let's look between 9 A.M. and 4 P.M. Monday through Friday — and *see* what happens to Kansas City radio listening when "all those teenagers" are at school. WHB *continues* its domination! According to every major survey, every one of the 140 quarter hours from 9 to 5 belongs overwhelmingly to WHB. This, mind you, when there are *no* teenagers available. No wonder WHB carries regular schedules for virtually every major Kansas City food chain — including A&P, Milgram's, Thriftway, A&G, Wolferman's and Kroger.

Sponsor, December 14, 1957: [At the top of the page there is a cartoon of an ancient radio with an old fashioned separate speaker on top of it. The caption reads:] In Kansas City, why settle for Podunk power? [A second cartoon at bottom of the page shows a very modern couple speaking into a modern microphone in a modern studio. Its caption reads:] Get 50,000 watt coverage on KCMO-Radio.

Sponsor, December 7, 1957: [Storz had already answered KCMO's jibe about WHB's "mere" 10,000 watts of power. In this advertisement, there is a photo of a completely empty stadium, with a businessman holding up a microphone ... to nobody. The caption reads:] COVERAGE? Yes ... but who's *listening*? It's WHB's 96-county world. IT'S A WHB PULSE! WHB is *first* in 432 of 432 quarter hours 6 A.M. to midnight.

One outcome of Storz's trade magazine advertising was that the Storz programming success story was in the consciousness of many of the broadcasters who subscribed to those publications. A significant number of them decided that it was time to make a visit to Kansas City, check into a hotel with a radio and a tape recorder, and start making airchecks of the Storz Sound so that they could go back home and attempt to put the format on their own station.

Gordon McLendon is one already successful multiple station owner who did just that. In effect, Storz's trades advertising helped to drive the dissemination of the Top 40 format to cities and towns across America. However, most imitators were not immediately able to fully reproduce the nuances of the Storz format, nor provide the quality of personnel, the committed management, and the multiple years of experience which the Mid-Continent Broadcasting Company brought to each market they entered.

The audience dominance which WHB achieved in a short span of time was truly remarkable. Other AM stations in Kansas City largely failed to mount a viable long-term challenge to WHB. It was not until the growing popularity of FM in the early 1970s that a station such as KEBQ-FM could begin to reduce WHB's market share by offering a high-energy format on the FM band. As FM listenership increased, WHB converted to playing adult contemporary music, then changed again to an all-oldies format, but WHB's audiences continued to drift away to other stations.

• FIVE •

Signals from the Frozen North, the F.C.C., and the Sunny South: WDGY and WQAM

In 1955, the growing popularity of Storz programming in Omaha, New Orleans, and Kansas City meant that there was money in the bank. Thus, at the end of the year, the company announced the purchase of WDGY, Minneapolis–St. Paul, for $334,000 (about $2,792,923 in 2012). Steve Labunski was moved from KOWH to become general manager, and Don Loughnane was assigned as WDGY's program director. He was soon replaced by young William L. "Bill" Armstrong, who must have needed a complete change of wardrobe to make the transition from steamy New Orleans to subarctic Minneapolis. (A comprehensive website at www.radiotapes.com/wdgy.html offers airchecks and photos of WDGY's facilities. Included is a master's thesis by Jerry Verne Haines, "The History of Radio Station WDGY" which provided valuable details of the station's early days.)

Like WHB, WDGY had a long history. The station had taken to the air (using different call letters) with a mere five watts of power in January of 1924.[1] It went through a bewildering number of changes in its studio and transmitter locations and transmitting frequency, until it finally settled at 1130 kHz in 1941. Between 1952 and 1954 — two years prior to Storz's takeover — WDGY had tried airing a current hit records format, but the programming was inconsistent. When the station began operating under Storz ownership in January of 1956, it was rated seventh in a seven station market.

With the purchase of 50,000-watt WDGY, Todd Storz had set his sights on what would be his largest market yet: the top 15 metropolitan area of Minneapolis–St. Paul. But unlike the dominance WHB had achieved so easily in

Kansas City, this time he would have to share the stage with other major stations. The biggest audiences belonged to the full-power regional signal of well-entrenched WCCO, a "full-service" station. "Full-service" stations typically have strong local and national newscasts, usually one or more local talk or call-in shows, and play-by-play coverage of the major sports franchises. If popular music is aired, it is likely to be conservative and aimed at a broad listenership. WCCO had all of those attributes. Later, WDGY would also have to battle against the Crowell-Collier company's personality-and promotion-driven Top 40 competitor KDWB.

On paper, WDGY's 50,000 watts of power — the maximum allowed for U.S. AM radio stations — meant that it was five times more powerful than WHB, and had 100 times the signal strength of KOWH. But WDGY shared its 1130 kHz frequency with stations elsewhere in the U.S., whereas WCCO was a "clear channel" station — no other entities transmitted on WCCO's frequency. To prevent causing interference, WDGY was required to operate a multiple tower array which narrowed and directionalized its signal. Maintaining that high-powered transmitter and complicated tower system would prove to be expensive.

Storz invaded Minneapolis with what had become his winning formula for success: the Top 40 format, heavy promotion, strong management, and top personalities. Some who remember WDGY in that initial period say it became his classiest, best-sounding property to date. But unlike the quick and continuing success that KOWH, WTIX, and WHB had enjoyed, WDGY never became the dominant signal in the market.

By the time Top 40 programming began on WDGY, Storz's tactic of saturating new stations with "listen-and-win" contests was well-known to other broadcasters. As reported in Haines's thesis, WTCN began giving away money even before Storz took over (p. 83). Haines produced a "Giveaway Box Score" that had appeared in the *Minneapolis Tribune* which showed the fierce level of competition WDGY had faced:

> WDGY gave a total of $2,940 to 49 winners, WTCN gave a total of $3,124 to 187 winners, KSTP gave a total of $4,670 to 38 winners, WCCO gave a total of $4,500 to 10 winners, and WLOL gave a total of $820 to 2 winners [p. 84].

Haines's recounting of the promotional "war" at its height nicely illustrates how far stations were willing to go to win:

> WCCO hired a special announcer to conduct its contests. "Big Bill Cash" was programmed specifically to fight WDGY's contest operations. WDGY retaliated by announcing it would "cooperate" with WCCO and rebroadcast WCCO's clues — hence there was no need to listen to WCCO. WCCO countered by changing its clues to items like "WCCO is tops," "3,000,000 Northwesterns listen to

WCCO," "I always listen to WCCO," and "More people listen to WCCO than to all other Twin Cities radio stations combined." Within one or two days, WDGY aired the announcement that it would no longer present WCCO's clues as WCCO had failed to give away the promised amount of money [p. 84].

In spite of the competition, WDGY moved into second place in the Hooper ratings after only five weeks of Storz programming. Haines reported that although the station had started at 4.2 mornings and 3.7 afternoons, after only three months of the Storz formula the ratings had jumped to 14.8 and 17.3 (p. 85).

In *Sponsor* magazine for May 14, 1956, the Mid-Continent Broadcasting Company ran a display ad titled "Much ado ... about something." The text read:

> Four Minneapolis–St. Paul radio stations, not fully satisfied WDGY was *really* in second place, hired a local market analyst to study the audience. WDGY wasn't expected to make a showing. WDGY wasn't invited to take part ... but WDGY turned up ... in *2nd* place. That's what Hooper said in the first place, and says again for March–April. Newest Area Nielsen shows WDGY gained 93 percent over the previous Nielsen audience share. All this just since February, when Mid-Continent news, music, and ideas came to Minneapolis–St. Paul ... and started rewriting the radio listening story.

After nine months, during which Jack Thayer replaced Steve Labunski as WDGY's general manager, the station could boast about another jump in ratings — to 22.8 mornings and 23.2 afternoons. Only WCCO had higher audience figures — but WCCO's number-one dominance would remain throughout Storz's ownership.

Mr. Haines described Storz programming on WDGY as "colorful" and provided two examples:

> Announcer Jack Thayer was put into a trance by a hypnotist. Once in the trance, Thayer proclaimed himself to be a Prussian army lieutenant. This took place during the "Bridey Murphy" [reincarnation] craze. DJ Stanley Mack repeatedly played the bizarre "Dinner with Drac" song one day and was "fired" on the air by Thayer, then general manager. Teenage outcry was so strong that WDGY was "forced" to rehire Mack and continue play of the record [p. 91].

Haines also offered this assessment of some of WDGY's air personalities:

> For its time, WDGY was notable for its degree of announcer personality ... Herb Oscar Anderson sang along with the records on his morning shift and interjected many personal observations. Bill Bennett was "the singing disc jockey." Dan Daniel, in the evening teenage hours talked about dating and the problems of acne (for the makers of acne medications) [p. 92].

Later in his thesis, Haines recounted how in 1960, Crowell-Collier — the new owners of KDWB at 630 kHz — began to give WDGY strong competition. That eventually brought both stations to a nominal ratings parity, where they continued to jockey for second or third place behind perennial leader WCCO.

One particular point stands out in Haines's "conclusions" section. He notes that in the years prior to Storz's takeover, the owners of WDGY had built the expensive-to-operate nine tower directional antenna system so that they could maximize the station's power

> as a public relations, image-building gesture. Any increased prestige derived from being a full-power station was far outweighed by the loss of audience [because of the highly-directional signal] and increased operating costs [incurred by the high-powered transmitter driving antennas on nine towers] [pp. 101–102].

It is worth noting that when KDWB began challenging WDGY with their version of Top 40 radio, they did so with a low-on-the-dial signal and a much smaller electricity bill.

In 1964, Mike Sigelman — fresh out of the University of Minnesota School of Journalism — interviewed for a job at WDGY in the cramped quarters of the Bloomington, Minnesota, transmitter site, which then also housed the on-air studios. "I had a picture in my mind of my first job in a swank office in a major broadcasting center," Sigelman remembers. For a while, WDGY's offices and studios had indeed been located in downtown Minneapolis on South Second Street, but in 1964 WDGY facilities out of town at the tower site were comprised of just a single on-air studio, a newsroom opposite the giant transmitter, and a production space in the basement. Even General Manager Dick Harris's office was small. But Sigelman wasn't disappointed.

> Dick was enthusiastic and I was immediately impressed by his quick mind and "driven" ways of making things happen. I was glad to accept his offer as Promotions Manager shortly after my arrival. I was reminded twice a day of WDGY's big power when they switched between 50,000 watts days and 25,000 watts nights. There was a loud BANG! of the antenna relays that would rock the building as we kicked the power up at sunrise or took it down at sunset. We could hardly hear the signal at night just 20 miles either way from the transmitter, but the signal boomed into northern Minnesota and lots of Canada. It sounded as if it were a local station in places like Brainerd, Minnesota — which was a few hundred miles to the north.

Sigelman remembers that the air staff spun their own records, but that WDGY was automated overnight when he arrived at the station in 1964. WDGY was loading all of its commercials into an automation system at that time, even when live talent was running the station — a forerunner of today's computer assisted automation. However, the station was on its way to returning to an "all-live" operation, and that is when the ratings started to return.

WDGY was dominant at night with Tall Paul Bunyan (Tom Campbell), a big personality who was the station's teen idol.

In regard to some of WDGY's promotions, Sigelman recalls,

> In 1964, the well-known St. Paul Winter Carnival featured the WDGY All Star Concert, with the hit music stars of the day. It was the beginning of the return to "the big time" for WDGY, as some 20,000 fans jammed the St. Paul Auditorium. That was also the year that [station manager] Dick Harris tried everything but could not land the Beatles to headline the show. The Beatles finally did schedule an appearance in Minneapolis a year later, giving rise to a battle between WDGY and KDWB over which station would host the Fab Four in their studios, and which would get to introduce them at Met Stadium in Bloomington.

In 1966, Dick Harris left the Storz organization and Phil Trammel replaced him as manager of WDGY. Sigelman said, "As I look back, I think he did little but cut overhead instead of building like Harris had done. That seemed to be the beginning of the end for the Mighty WeeGee."

But the cost cutting actions that Sigelman attributed to Trammel, and later to Trammel's replacement Dale Weber, weren't taken under their own volition. They were the financial philosophy of Todd's father, Robert H. Storz, who — following Todd's premature death — ran the stations like branches of the banks that Robert understood, not the entertainment outlets that his son had perfected. Sigelman concludes:

> When WDGY was sold to Malrite and — out of desperation — changed to a country format under the leadership of Weber, the turnover of programming staff accelerated. The format changeover was heralded by a television commercial that showed a cowboy lassoing a hard-rock artist to the ground and riding off into the sunset with his guitar — leaving the legendary rocker that had been WDGY grounded forever. Ten years after that, country music listeners moved to FM, and the AM 1130 frequency changed to all-sports as KFAN. With even the call letters gone, the "Wonderful Weegie" story had reached its end.

Later, Mike Sigelman became Deane Johnson's sales manager, in the period from 1969 to 1972 when Johnson was both manager and program director of WDGY's fierce Top 40 competitor, KDWB.

Deane Johnson picks up the WDGY story in late February of 1969, when he was contacted by Grahame Richards, a former Storz national program director.

> We had remained close friends since our Storz days. At that time he [Richards] was consultant for the owners of KDWB in Minneapolis–St. Paul and needed to recruit a program director. I flew to Minneapolis to meet with the owners and received an offer. Upon returning to Omaha, I decided to decline the opportunity and remain as program director of KOIL in Omaha [which had been chief competitor to Storz's KOWH]. Friends in the industry advised me

Five • The Frozen North, the F.C.C., the Sunny South

against getting involved in the KDWB "meat grinder" and it did seem to be a hopeless situation. But after two increased offers, the money was substantial, and I decided to risk going to Minneapolis and taking on Storz's WDGY. Little did I know at that time that I would later be appointed general manager and would serve as both GM and PD simultaneously.

Upon arrival, and after a few days of monitoring WDGY, I realized that this Storz station had a good staff, was consistently smooth — running a modified Drake [more music/less talk] format with 20/20 news — but mixing in more personality than the Drake format. It was a typical Storz-programmed station, but much tighter than the Storz format of earlier days. WDGY had momentum. KDWB, on the other hand, was a pit. As I recall, only one member of the talent staff was viable against WDGY.

In those days, a strong evening personality was a necessity, and evenings were the first place one could make rating gains, which could then be spread into daytime hours. I realized that WDGY had an excellent evening jock in Rob Sherwood, who would be tough to beat. I needed to weaken WDGY by bringing him over to KDWB.

I called Sherwood one evening while he was on the air, explained who I was and that I thought he was a great talent, and that I'd like to talk with him about moving over and helping me against WDGY. Surprisingly, he agreed to meet. He declined the first offer, and he also passed on a subsequent offer. He was a successful evening personality on a 50,000-watt Storz station with excellent ratings; why should he move to a "pit"?

To break things loose, I offered him an all-expense-paid trip around the world as a signing bonus. That did it. He moved to KDWB, after asking, "Are you going to straighten up the rest of that mess over there?" Ironically, Sherwood never got around to taking the around-the-world trip, but later traveled to England to report for KDWB on "Beatleland."

Just when I thought I had pulled off a coup with the Sherwood deal, WDGY countered my move by bringing back a previous evening jock, in effect replacing Rob Sherwood with an equal act with whom listeners were familiar. It was obvious WDGY was not going to be an easy takedown in the ratings war.

Regardless, we whittled away at WDGY. Finally, in a [1972] Arbitron rating taken three years after I arrived — the last Arbitron before I left the station — KDWB topped WDGY in every category by a small margin. The battle against Storz's WDGY had been won, but it was a narrow victory — and a short-lived one.

The tables were about to turn once again in WDGY's favor. The ownership of KDWB had decided to exit the broadcasting business while KDWB was on a roll, and sold the station to Doubleday, the book publisher. They would be bringing in their own management team, which is why I left the station, even though I had been asked to continue programming it. Once again, new ownership fumbled the ball — especially with poor management — which resulted in their giving a substantial ratings advantage back to WDGY.

But the winds of change were blowing in the market by 1972. KQRS, an Album Oriented Rock [AOR] station was poised to make significant gains, further dividing the contemporary music audience. KDWB under Doubleday ownership eventually acquired an FM, something the Storz company under Robert

H. Storz had foolishly declined to do. The heyday of Storz market domination with the Top 40 format on AM would gradually subside in Minneapolis–St. Paul, just as it did during the 1970s in other Storz markets.[2]

Mike Sigelman and Deane Johnson's descriptions of how WDGY struggled to maintain its initial rise to first place in the Minneapolis–St. Paul market, and how it eventually faded to near irrelevance as FM audiences grew and music listening choices proliferated, would become the unfortunate pattern for all of the Storz stations.

That unhappy demise could not have been foreseen by Storz back in the spring of 1956, when all of his stations were profitable and Top 40 was all the rage. In May — just six months after purchasing WDGY — Storz bought his fifth station: WQAM in Miami. It would be the toughest acquisition Storz ever encountered from a legal standpoint, but perhaps the most important one for him from a personal point of view. For Storz, it would be a chance for fun in the sun at last.

"Todd loved New Orleans, but he loved Miami more," recalled Ruthie Petersen, Todd Storz's secretary. South Florida in 1956 seemed to be a glittering jewel of new fashion, avant-garde construction and welcoming sunshine to thousands of Florida newcomers. In many ways, Miami was the synthesis of New York new money and Los Angeles glitz. Within a year, Storz would navigate through a difficult station acquisition, purchase a building on Miami Beach, and move the home office there from Omaha. Some believe he moved to Florida for his health or for better weather, but most concur that he wanted to escape the daily scrutiny of his father, the senior and principal investor in the corporation.

On May 14, 1956, *Broadcasting-Telecasting* ran an abbreviated item reporting that the *Miami Herald* newspaper (owned by the Knight newspaper and radio interests) had sold WQAM, Florida's oldest radio station, to Storz's Mid-Continent Broadcasting Company "for what is believed to be [a] record price for [a] regional of $850,000 cash" (about $7,092,968 in 2012). Putting the station up for sale had been forced by the Federal Communications Commission's "duopoly" policy then in effect, which stated that a company could not own more than one station on the same broadcast band (AM or FM or TV) in the same market. In this case, a license for a new Miami television channel had been awarded to a corporation which had ended up owning both WIOD-AM and WQAM-AM — violating the FCC's "only one of each" rule.

One month after announcing the sale of WQAM on June 4, 1956, *Time* magazine published a sarcastic and negative article about Todd Storz and his growing roster of stations, titling it — and him — as the "King of Giveaway." It harshly critiqued Storz's methods of attracting audience. The first line of the article began: "Todd Storz, 32, whose low estimate of listeners' intelligence is tempered only by his high regard for their cupidity." Under Todd's photo was the caption: "He Stops Traffic." The article included an admonition attributed to Kansas City Chief of Police Bernard Branson declaring "the past-time [i.e., treasure hunts] should be banned." The stations may not have been operating in "the public interest, convenience, or necessity"—a phrase from the 1934 Communications Act applied by the FCC in its licensee oversight. Bob Murphy, writing in the *Minneapolis Star*, reported that Storz had told the commission "he was unaware that the FCC looked with displeasure on such [promotional] programs." The FCC would later acknowledge that the *Time* article had raised public interest questions about Mid-Continent Broadcasting's operations, and thus the future sale of Miami station WQAM to the

Todd in 1956, closing the deal to acquire WQAM from James L. Knight, who also owned the Miami Herald newspaper. That purchase became problematical when the Federal Communications Commission questioned Storz's pattern of "buying" audiences with giveaway contests. The implication was that Storz's stations were not operating "in the public interest" and therefore their licenses might not be renewed.

company. *Variety* (July 17, 1956) reported that Storz had been given thirty days to show why an FCC hearing into the proposed purchase should not be held. The commission had sent Storz a so-called "McFarland letter" advising him that instead of rubber-stamping the sale, a hearing would be needed. The McFarland letter was an FCC form letter which usually advised an applicant for a station license that his bid was mutually exclusive with that of another applicant, and which asked the applicant to submit an explanation or more information. The FCC's letter to Storz went beyond that. It stated that KOWH and WDGY's recent treasure hunts indicated that those stations were "purchasing" the listening audience, which had prompted competing stations to try the same thing as a counter-measure. The FCC added that "this pattern of operation, with its apparent success, appears to be an inducement to other broadcasters to adopt similar methods" which "results in a deterioration in the quality of the service previously rendered the public." Many broadcasters of the day felt that this was an unprecedented intrusion into programming by the FCC.

Storz made at least two replies to the McFarland letter. In the first, he pointed out that contests and other giveaways accounted for only one percent of airtime and were thus an insignificant programming factor, and that hundreds of other stations were doing the same thing.

His second letter, sent on July 18, was more conciliatory. The commission was scheduled to recess for the summer after its July 19 hearing, which would mean that no action would be taken on the WQAM transfer until after Storz's contract to buy the station expired on August 15. So in the second letter, Storz made this promise via a notarized affidavit:

> Mid-Continent Broadcasting Co., will, upon the commission taking favorable action on the instant application involving stations WQAM and WQAM-FM, discontinue all contests and/or "giveaway" programs designed to attract audience or influence listening over all broadcast facilities owned by it as soon as possible.

It was a bold but also an apparently necessary move. Storz had promised to drop all contests and giveaways at all of his stations, not just WQAM. But as it turned out, the FCC did not base its decision on the promise Storz made in his second letter, but on the much broader question of whether or not the commission had the right to consider programming in any of its decision-making. The commissioners voted four to three that it did not, resulting in Storz getting the WQAM license.

In a July 23, 1956, memorandum, Storz explained the situation to his colleagues:

> I have just completed a very trying ten days in Washington on the WQAM transfer. The question ... was not whether the commission would approve the

transfer to us, but instead, whether the commission would approve the transfer to us without a hearing.... If the commission had failed to act, or if the commission had voted to send the matter to hearing, we would have, in all likelihood, lost the purchase of WQAM.... The truth of the matter is that we discontinued these contests because they were questioned by the Federal Communications Commission.

The specter of censorship by the FCC had become a big topic of conversation among many broadcasters, but especially TV moguls. Storz further observed: "if the commission eventually determines that large money contests are not in the public interest, we feel certain it would have to discontinue all such contests whether originated by a local radio station, or by some other source such as a TV network."

The latter remark pointed to the hypocrisy of the FCC allowing major TV networks to air big-money quiz shows while looking down their noses at the Storz stations' giveaways. But the implicit message was this: Radio programming content decisions could adversely affect the FCC licensing process. The WQAM controversy had a lasting impact on both the Storz stations and the entire radio business.

To comply with Storz's promise to the FCC, program directors informed their announcers and continuity directors (who prepared the daily radio station program logs) to phase out and discontinue all existing cash-based stunts and promotions. A memo issued on July 21, 1956, by newly named KOWH Program Director Bill Stewart said in part:

> As soon as we have our next LUCKY HOUSE NUMBER WINNER, it will be discontinued with no mention whatsoever on the air as to why it was discontinued.... The CASH FOR KIDS promotion will taper off ... AUTO CASH will cease almost immediately. The important thing to remember is ... ABSOLUTELY NO MENTION ON THE AIR ABOUT THE CONTESTS BEING OFF!

Storz's decision to end the station stunts and promotions which had helped to make him rich and famous became the talk of the radio business in the summer and fall of 1956. By the end of the year, he was ready to make a statement to the broadcast industry that would set the record straight. While it put the lid on the Storz stations' most visible promotions, the Storz stations' new promotional restraint was not emulated by others in the industry.

A typewritten letter to the editor ran as a display ad in *Advertising Age* (December 31, 1956). It was titled "Programs Include Many Services, Storz Head Says."

> To the Editor: First of all, may I commend you for devoting a considerable portion of your Dec. 10 issue to the broadcasting industry, and thank you for the references you made to the Storz stations. At the same time, we seriously object

to being described as "juke box" stations, a phrase invented by competitors of successful independent radio stations offering modern, entertaining and informative listening which today's radio audiences have found much to their liking. It is no more apt a description of us than if we referred to competitive stations using network programming as "soap opera machines" or "do nothing specialists."

Furthermore, it is simply not true that we indulged in "an orgy of audience buying ... in Minneapolis." It is a fact that in what we felt was a proper ratio to the overall broadcast day, we engaged in several contest and so-called "give-away" programs simply as a small part of a very much larger and continuing promotion of our station WDGY in Minneapolis. It is particularly significant that we have long since discontinued all such "give-away" programs at all of our stations, and our audiences have continued to grow everywhere. The most revealing and most recent example of this success has been in Miami, where WQAM vaulted into first place in audience in approximately 90 days without the use of a single so-called "give-away" contest. The theory that contests "buy audiences" has been exploded once and for all by our WQAM experience. As a further interesting footnote, some of our competitors in Minneapolis and the other cities are still running various "give-away" contests and are generally not succeeding in gaining audience.

We are convinced that the size of a station's audience closely parallels the entertainment value of the programming content offered and the over-all service provided to the station's listening area. As an organization, we concentrate much of our time and energy in trying to assure maximum audience appeal in everything we do.

As you undoubtedly know, the programming content of our stations includes many features and services other than the playing of popular music. These include a large number of newscasts and special events, carefully selected religious programs, some agricultural features, a private weather service, as well as numerous other public service features of every type and description. Our overall programming is designed not only to attract and entertain, but to inform and to serve our vast and growing listening audiences.

At times, it seems that nobody likes our programming but the listeners.

Todd Storz, President, The Storz Stations, Omaha

By the time this advertisement was published, listeners in South Florida and the nearby Caribbean were beginning to be counted among those "nobodys" who liked Storz programming. WQAM occupied an inherently stronger low-end-of-the-AM dial position at 560 kHz, and its transmitting tower was grounded in the salt water of Biscayne Bay, together providing the optimal physical factors for generating a powerful signal. WQAM could be heard everywhere within hundreds of miles of Miami. In addition to completely covering south Florida, the station beamed a strong signal over much of the Caribbean and all of pre–Castro Cuba, twenty-four-hours a day. Under Storz ownership, WQAM advertised the USA's best-known products, South Florida's major businesses, and the extravaganzas featuring top show business personalities appearing nightly in Miami Beach's glamorous high-rise hotels.

Five • The Frozen North, the F.C.C., the Sunny South

In August of 1956, *Broadcasting-Telecasting* announced that Jack Sandler — who had begun his radio career at KOWH as a sportscaster — had been named WQAM's general manager. A little less than a year later, in April of 1957, *Billboard* magazine announced that "Kent Burkhart, who has been in radio since the age of 10, has joined WQAM, Miami, as Program Director." Burkhart would become nationally famous for his programming savvy. Later, he was a distributor of programming services via satellite, and half of the highly successful Burkhart/Abrams radio programming consultancy.

WQAM's Bobby Lyons, a transfer from KOWH Omaha, was Storz's voice of choice for many of the live commercials for Miami Beach hotel shows and Havana's swinging nightlife, which was a mere hour from Miami by an inexpensive shuttle flight. Lyons's delivery and his comedic alternate voices made him one of Storz's best performers on "The New WQAM." For one thing, Lyons could do a perfect imitation of CBS's talk host Arthur Godfrey, sometimes calling an unsuspecting advertiser, media personality, WQAM listener, or political figure with his Godfrey routine. According to Bud Connell, he could do an even better "put-on" with his Donald Duck voice.

In late 1956 and early 1957, Bill Stewart was serving as program director at KOWH. In a 1971 interview, he recalled that although the early WQAM referred publicly to its programming as being the Storz Top 40 format, "when we [Storz Broadcasting] first went into Miami ... we went in with I think 25 records. I don't think it ever got over really 30 records. We would *print* a Top 40, but we only played 30. And it worked. Because ... it eliminated chance, that's the only thing it did." The stratagem wasn't uncommon on stations that were encountering strong "current hits" competition, but Stewart's admission to using the practice in late 1956 is remarkable because that was before the time when Top 40 outlets were saturating the airwaves. Bud Connell, who was a KOWH air personality during this same period, argues that Bill Stewart could have applied the shortened playlist on WQAM during a sliver of time in 1956, but that would be all: "When I first joined KOWH, we played the entire Top 40, 6 to 8 extras, plus a Pick Hit, a total of 47 to 49 songs. I am not aware of any edicts to the other stations restricting airplay to a certain number of records."

The glamour of being a young Top 40 disc jockey in a vibrant city like Miami must have been intoxicating. But WQAM's teen-appeal personality Rick Shaw recalled, "When I met Todd for the first time, we were talking radio, and I asked him to describe for me my role. He looked at me, thought about it, and said, 'Your job, your role on my radio station, is to create the most positive kind of environment for the delivery of a commercial message.' I said, whoa! That's it! One sentence. That pretty much covers it."

Todd Storz, executives and managers, 1957. This photograph was likely taken in Miami, home of WQAM. In the front row (left to right) are Todd Storz, Bud Armstrong, WQAM manager Jack Sandler, and Dale Moudy. In the back row are Fred Berthelson, Steve Labunski, Bill Stewart, and Virgil Sharpe. Stewart seems to be looking at a reflection from Jack Sandler's bald head.

In 1957, fresh from a tour of army duty in Korea and then a job at a Richmond, Virginia, radio station, Charlie Murdock paid a visit to WQAM studios during a Florida vacation and made an audition tape. After listening to Charlie's tape, Kent Burkhart offered him a job on the spot, placing him in the important afternoon drive slot. Murdock became program director in 1958 and was named operations manager four years later.[3]

WQAM's rise to dominance in South Florida was typical of what had happened in the other markets Storz had entered, even without the usual giveaways and prize contests. The front cover of the December 3, 1956, *Broadcasting-Telecasting* carried a full-page ad with the headline: "WQAM leaps to 1st place in audience after less than 3 months of Storz programming." A box reporting the all-day Hooper ratings for competing stations showed WQAM at 28 percent and the next station — "Station A" — at 18 percent. Just one week

Five • The Frozen North, the F.C.C., the Sunny South

after that, the magazine carried a two-page Storz ad inside. It appeared in the form of a "memo" from Todd Storz on "Storz Stations" letterhead.

The memo was addressed to "Jack Sandler, General Manager; and ALL THE STAFF AT WQAM," and carried Storz's signature at the bottom. This kind of advertising was a departure for Storz, who rarely allowed himself to be the focus of attention. The text of the letter was as follows:

> I want to offer my heartiest congratulations to you on the newest Hooper survey for Miami which covers the months of October–November, 1956. As you know, it shows WQAM in first place. *First Place* in the morning, 7:00 A.M. to 12:00 noon. *First Place* in the afternoon, 12:00 noon to 6:00 P.M., and, of course, *first place* in all-day average.
>
> This has been accomplished in just a little over 90 days. Frankly, I can hardly express in words the pleasure and satisfaction this great achievement gives me.
>
> To see just how great an accomplishment it is, let's look at it in the light of history. The objective of Storz Station programming has always been to provide the people served by our stations with programs of maximum interest and entertainment value. Pursuing this basic objective, each of our stations became — and remained — the most-listened-to station in its area, according to numerous surveys and audience reports.
>
> In the past we also broadcast several contests and "give-away" programs, feeling that in proper ratio to the overall broadcast day, these, too, had interest and entertainment value for our listeners. However, it was always our contention that contests and give-aways of themselves could not, and would not, build and maintain station audience. Unless overall station programming philosophy were sound, contests would add little, if anything, to the audience.
>
> Shortly before our company took over the operation of WQAM, we became aware of information leading us to conclude that the Federal Communications Commission frowned on the broadcasting of contests and give-aways. Immediately, on all of our existing stations, we discontinued broadcasting such features. Under our ownership, WQAM, as you know, has never broadcast any give-away, or any contests requiring the listener to be tuned in in order to win a prize.
>
> This fact itself underscores the fabulous job done by WQAM. The credit for this achievement goes justly to every member of the WQAM staff. The tremendous and rapid growth of WQAM to a position of first place dominance is a direct result of the enthusiasm and dispatch with which you have executed the creative ideas. I know of no parallel anywhere in the radio industry. Again, my hearty congratulations for a difficult job, well done.
>
> Our present audience position is reassuring, with an all day average of almost 30 percent, while the second station has 18 percent. But we are hopeful that this is only the beginning. Greater Miami, now grown to a population in excess of one million, is entitled to the best radio that our ability, interest and creative effort can produce.
>
> The programming committee has now developed 34 new programming ideas, which will be put into effect on WQAM as quickly as possible. Some will go on the air immediately. All should be in effect by next Spring. [Signed] Todd Storz.

The "programming committee" that had "developed 34 new programming ideas" was probably fictitious, but the reader did get the desired impression that — in the vernacular of the day — "you ain't seen nuthin' yet!" Less than two months later, an ad in *Sponsor* for January 19, 1957, trumpeted "in Miami ... WQAM has made it even more of a runaway ... without a give-away!" That became the key phrase in WQAM trades advertising for about the next nine months. By May, another *Sponsor* ad bragged that "WQAM nets more than twice the daytime audience of the runner-up station. All three agree: First it was Hooper ... then Trendex ... now Pulse shows WQAM in First Place." A bar graph showed a long horizontal line with WQAM at 34.1 percent, and station "A" at 15.6 percent. In July, Hooper showed WQAM at 37.9 percent and station "A" dropping down to 10.1 percent. By WQAM's first anniversary under Storz ownership, in October of 1957, WQAM's all-day average, 7 A.M. to 6 P.M. Monday through Saturday, was 42.1 percent. The ad in that week's *Broadcasting-Telecasting* headlined "First Anniversary as a Storz Station Finds WQAM's Runaway Complete!"

Inevitably, other stations stepped up their efforts to compete with the dominant WQAM. One was a new daytime-only station, WAME, which transmitted with 5,000 watts of power as WQAM did, but was positioned much farther up the AM band at 1260 kHz. In spite of those two handicaps, WAME mounted an aggressive advertising campaign in the radio trade magazines that parodied the look of Storz's adds, boldly declaring that they were "the new number two station in the market." In fact, WAME's catch phrase was "Whammy — Radio Two in Miami." They claimed to be the "dominant" number two station with 17.8 percent of the audience in October of 1959. WAME held on into the 1960s as a Top 40 station, but ultimately did not survive the battle about to be waged between WQAM and a powerful new contender.

Radio station group owner Robert Rounsaville had gained major market experience in Atlanta with WQXI, competing successfully with other music-and-news outlets there. He owned six other stations, including WMBM in Miami Beach, a weak 1,000-watt daytime-only facility operating at 800 kHz, which featured Black programming and rhythm and blues music. Rounsaville packaged the WMBM call letters along with his well-established Black community advertising business, and offered it to another Miami operator in exchange for nearly enough cash to build a new full-time facility at 790 kHz, transmitting with 5,000 watts days, and running a directional pattern at night. This new station would *not* be another high-end-of-the-dial, low-wattage WAME.

In the early fall of 1960, Rounsaville contacted a young man who was relatively well-known for being the first programmer to take the top ratings spot from a Storz station. That man was Storz's former KOWH disc jockey Bud Connell, who had consistently bested Storz's WTIX in New Orleans for nearly three consecutive years. Connell was already looking to make a change, and a move to Miami to program against another Storz station would be a perfect fit. Rounsaville told Connell that he wanted to build a "general market" station to appeal to the same audience that Storz's WQAM controlled. Rounsaville negotiated a management deal with the understanding that Connell would become station manager and also control the programming of the Miami project beginning November 1, 1960.

One of the first puzzles to be solved was what the new station's call letters should be. Connell provided Rounsaville with a list of available calls, plus several which might be available for a price consideration. Among the latter group was WFUN, a call assigned to a small-market station in Alabama. Rounsaville loved the idea, made a quiet middle-five-figure offer, and the Alabama station dropped WFUN as its call letters, leaving Rounsaville's legal representatives clear to apply for them the next morning. The new 790 in Miami would be "WFUN — Fun Radio" with a target sign-on date shortly after the first of the new year, 1961.

In the waning days of 1960, Connell mounted a ninety-day hiring, creative writing, recording, and rehearsal for "Fun in the Sun" radio. Following two weeks of 24-hour, off-air, full broadcast simulation, he signed on his new creation in late January of 1961. Todd Storz, who had rented a suite at a major hotel to monitor WFUN's opening, listened to the first two hours and reportedly told other Storz executives assembled in the room, "It's all over."

Within another ninety days, the polished new competitor rivaled WQAM's ratings in most of the broadcast day, and became the most talked about station in Florida. When one promotion ended, another — which had been carefully researched, prepared, and pre-recorded for broadcast — hit the airwaves.

WFUN's 790-pound "Trunkful of Money" went on display in a guarded window of Burdine's downtown department store — the first person who guessed the exact amount that was in the trunk would win it all. A long list of "exotic commercials" hit the airwaves, including an advertisement for King Burgers in Seattle (for which Connell was threatened suit by Burger King), another for the Brooklyn Staten Island Ferry, still another offering a "very much alive Atlantic ocean shark for your swimming pool." One of the faux commercials was taken quite seriously when WFUN was "proud to announce" the offering of the elegant diamond on the beach — the Fountainebleau Hotel and its surrounding acreage — available for the sum of only $35,000,000, which was a vast fortune in 1961 (equivalent to about $265,315,217 in 2012).

A potential buyer actually stepped forward to negotiate. Plastic eggs, stuffed with cash, candy, and certificates for prizes, were deposited on 50,000 residential lawns overnight. A deejay was fired on the air for merely mentioning WQAM's call letters, and later, a statewide manhunt ensued to get him back on the air "due to public demand." One promotional assault after another prevented WQAM from reclaiming its first place mantle for several years.[4]

Bud Connell remembers that WFUN "salted" their trash cans with "first drafts" of phony upcoming promotions because he knew that WQAM staffers were looking for clues to the upstart station's next big programming attack. Says Connell, "Of course, the phony promotions were snarled messes that were impossible to decipher, and we had the police stake out our trash. They caught WQAM personnel red-handed."

According to WFUN owner Robert W. Rounsaville, the station was always "neck and neck" with WQAM in the Hooper ratings, and in its early days was ahead of WQAM for twenty-two months. During the mid–and-late-1960s, he said, "We haven't beaten them decisively nor have they us. It's been a 'Mexican Standoff.' When we've beaten them it's been by decimal points, and they've beaten us the same way." (You can listen to a montage of 1960s WFUN airchecks at *www.wfuntribute.com*. Scroll down until you see "Airchecks," then in red letters "WFUN AUDIO." Click it. That takes you to a similar screen. Click on "02.WFUN 1960s Montage.")

Just three months after the beginning of WFUN's attack on WQAM's dominance, a very different kind of assault was being prepared by no less than the Central Intelligence Agency. And they were requesting help from WQAM.

One afternoon in mid–April 1961, Charlie Murdock approached the control room in WQAM's studios on the top floor of Miami's DuPont Building to prepare his deejay show for broadcast. Chief Engineer Morrie Barwick, who was usually busy maintaining equipment, had set up a large table in the control room for special guests. Seated behind the table were two stern, unsmiling dark-haired gentlemen in their middle to late thirties. Jack Sandler, WQAM's general manager, caught Murdock by the arm before he entered the room and quietly revealed that the two men were from the CIA, and would be recording a coded message to be broadcast several times throughout Murdock's afternoon show and again overnight. These were ostensibly to be heard, Charlie surmised, by 1,300 Cuban exiles and their Cuban supporters, all armed with U.S. weapons and waiting for instructions and timing for an invasion of the Bay of Pigs (Bahía de Cochinos) on Cuba's southern coast. Murdock pressed the red "record" button and watched the well-rehearsed men as they methodically laid out instructions in Spanish. They were confident and assured, as if they owned the radio station, giving what Murdock presumed to be coded

Five • The Frozen North, the F.C.C., the Sunny South

directives, dictates, orders, and timing. After recording the segment lasting approximately five minutes, the CIA agents directed Murdock to play the message back, probably so they could verify accuracy and completeness. The tape recording was then transferred to disk, labeled, and moved to the on-air studio for playback at prescribed times. Each playing was to be preceded by a carefully written introduction.

That first recording, as Murdock recalled, was introduced as a special message to "our Cuban friends in the Keys and elsewhere in the Caribbean," and was played just before Murdock's own 3 P.M. show time. Subsequent airings were heard later that afternoon, and overnight on WQAM's Al Martinez Show. The men from the CIA made clear that the message had to be broadcast precisely as recorded, and at specific times.

Charlie Murdock, center, with Cuban and U.S. government official. The identity of Charlie Murdock is not in question, but that of the other two men in the photograph is unknown. The man on the left could be a WQAM engineer, or from the C.I.A. The man at the typewriter could be a WQAM employee, or a reporter — or from the C.I.A. Initially, the entire operation to broadcast messages to the insurgents desiring to retake Cuba from Fidel Castro was a big secret.

The exiles who were supposed to hear this message were intended to cross the island to Havana with support from the local population. However, it became evident from the first hours of conflict that the exiles would likely lose the battle. President Kennedy then decided against his option of using the U.S. Air Force for support, and as a result, Castro's army quickly halted the invasion. When the fighting ended on April 19, 1961, ninety exiles had been killed and the rest taken as prisoners.

For six weeks before, during, and after the Bay of Pigs operation, WQAM's post-midnight programming included live Spanish-language programs which were fifteen to forty-five minutes in length. The programs originated from The Voice of America.

During that same tense period, WQAM's new competitor—WFUN—hosted the president of Free Cuba in their Miami Beach studios, as part of his effort to reach and encourage his people. He wanted to assure them that freedom would someday return to their beloved island of Cuba, and to the Isla de la Juventud, the Isle of Youth. WFUN sent a relatively strong signal over Cuba during its evening hours, except when Cuba "jammed" (interfered with) the signal on Fidel Castro's orders. According to Bud Connell, then station manager of WFUN, their chief engineer, Paul Cram, had been commissioned to build a transmitter capable of jamming *Castro's* signals!

Fairly quickly, the Cuban Communist revolution and the closure of Havana's glitzy hotels, casinos and nightclubs, stunted an important stream of advertising revenue for all of south Florida's broadcast outlets. Miami Beach's big hotels and major celebrity shows did expand their advertising to capture the former Havana tourist traffic, but not enough to make up for WQAM's loss.

Charlie Murdock had joined WQAM in August of 1957 as the station's afternoon deejay and sports director. He became program director in 1958, and operations manager in 1962. A few years later, he married WQAM receptionist Cecile Kirby and left South Florida and the Storz organization, becoming vice president and general manager of Cincinnati's legendary WLW in 1967. An aircheck of Charlie Murdock's afternoon show on WQAM from April 1962, can be listened to on *www.560.com*. The website also offers a range of other Storz-era WQAM airchecks, jingles and newscasts.

• SIX •

Programming Conventions I: Learning the Basics

In January of 1958, Todd Storz seemed to have a sunny outlook on the state of the Storz Stations as conveyed in a letter to his lifetime friend and former director of engineering Dale Moudy. The letter was on the Storz Stations letterhead, which carried the home office address of 222 South 15th Street, Omaha 2, Nebraska. In a small font on the top left were the words "Todd Storz — President," balanced on the top right by the current roster of stations: WDGY, WHB, WTIX, and WQAM — also in a small font. Moudy was then living at 7 West 66th Street in New York, where he had moved to work for the American Broadcasting Network — formerly ABC. ABN had attempted to compete against Top 40 stations with expensive live bands and singers. The network had lured Moudy to join the new venture by promising a much larger salary.

Storz often wrote business letters in a clipped and formal style, but this one has the tone of one family member "catching up" another family member on what has transpired lately. The "Fred" mentioned in connection with the purchase of WWEZ was Fred Berthelson, manager of WTIX. When Storz commented that "the negotiations and skirmishing on this deal have been going on a long time," he wasn't exaggerating — it had taken four years to acquire the better facility in New Orleans. "Cullum" was an engineer at WDGY who had not been able to completely tame that station's complicated nine-tower transmission system. Storz concludes the letter the way he began it — by expressing delight at the functional and "moderately luxurious" new studios at WQAM in Miami. Soon, Storz would move to that city himself, partly to escape his father's scrutiny, partly to enjoy the better weather, and partly to seek new companionship as his marriage headed for failure.

Thanks for your letter. Everything is going along fine, although we have been awfully busy the last few months.

The Miami studios turned out to be absolutely terrific. By far the most attractive of any we have, I think. It is a very functional, although a moderately luxurious layout. It has worked out very well and I'm sure it has been partially responsible for the much improved morale down there.

We have purchased WWEZ in New Orleans. This is a darn good facility with 5 kw on 690, and will give Fred a lot more secure position and a much better future. Actually, we will change the call letters to WTIX, and keep our present studios. No change in personnel is involved, except for our inheritance of the WWEZ transmitter engineers. The station operates with directional day and night, four towers in the line on about 42 acres of very expensive though wet ground just east of the city. The negotiations and skirmishing on this deal have been going on a long time, although the contract has finally been signed now.

Cullum is still fooling around with the antenna at WDGY, trying different phasing adjustments to get the pattern in tolerance to the northwest. The lawsuit has never come to trial since we had it held over to the next court session which will be in March.

The departure of employees has finally come to a stop, except for the recent disappearance of Chuckie from the Miami scene. Our few remaining bachelors are rapidly disappearing. Charlie Murdock is marrying Cecile Kirby, the receptionist in Miami, and Jim Ramsberg recently returned to WDGY from the service and has also gotten married.

I am sincerely sorry to hear about the oncoming demise of the new ABN. But I guess we agreed it was an extremely long shot from the beginning.

If you're out this way, stop in; or if you get down to Miami be sure to stop up and see the studios. The basic plan was really your handiwork and I know you would enjoy seeing them. Henry End did a really outstanding job on the decorations.

With best wishes, I am
Cordially yours,
Todd Storz

[In Todd's handwriting] P.S. Hello to Ellie & kids

At the time of Todd's letter in early 1958, the Storz stations were all doing well — even under-powered WTIX. KOWH had been sold for the highest price ever paid for a daytime-only station. Broadcasters across the nation wanted to learn the secrets of Storz stations' success, and thus was born a plan for an annual convention for disc jockeys and radio programmers that would include informative sessions about programming and procedures. The Pop Music Dick Jockey Convention and Radio Programming Seminars sponsored by Storz Broadcasting in 1958 and 1959 were landmark events for the company, but in very divergent ways. The first was a roaring success, but the second was a huge embarrassment.

The 1958 convention, hosted by WHB and held at Kansas City's

Muehlebach Hotel, March 7–9, 1958, set the intended tone of saluting the achievements of disc jockeys in building audiences, raising sales revenue, and becoming responsible figureheads in their communities. The event also offered forums intended to improve deejays' effectiveness and stature.

WHB general manager Bud Armstrong's "welcome" message, printed in the convention brochure, lauded the American disc jockey for being "as much an integral part of our national scene as hot dogs, ham and eggs, or the right of free speech. You disc jockeys are many wonderful things, to business men, housewives, old and young alike. You are friend and entertainer, companion and public servant. WHB and all the Storz Stations salute you and welcome you to this, your first convention."

Dale Moudy, about 1955. Todd Storz enjoyed a lifelong friendship with Dale Moudy. As Todd's director of engineering, Moudy built proprietary electronic equipment which made Storz stations sound different from competitors technically — and also in the sense that the stations had more programming "tools" available.

Registration to attend the events was free, and was "open to all Disc-Jockeys, Program Directors, Record Industry Management Personnel and Broadcasting Management Personnel," said a full-page advertisement in *Broadcasting* magazine for January 27, 1958.

The convention and programming seminars were Bill Stewart's project. That the initial event was praiseworthy, and that the second brought shame to the Storz stations, can be seen as a reflection of both Stewart's abilities and his failings.

The word that best describes Stewart's career in programming radio stations for various station groups and individuals is "peripatetic," which means "moving or traveling from place to place." Whether by his choice, or the decision of his employers, he did not remain long-term with any station or group owner. But such frequent relocation meant that Stewart was a "pollinator"—ideas that had just proved themselves at station "X" were soon germinating among stations in group "Y." When Stewart was not the originator of a new programming element, he often applied a "show business" sensibility that helped something prosaic seem remarkable, and claimed the result as his own new creation.

At one time, Stewart had the title of National Program Director for the Mid-Continent/Storz Broadcasting Company. At other times, he also held a similar title with the McLendon Stations, based initially in Texas. McLendon and Storz did not go head-to-head in the same markets, but Gordon McLendon's father-in-law, James A. Noe, owned WNOE in New Orleans, and Stewart agreed to program that station — against Storz's WTIX. (Bud Connell held the same post at WNOE and counter-programmed WTIX even more successfully.)

It is not possible to decide if Stewart had a loyalty problem, if he was merely on the lookout for new challenges, or was after more money. Perhaps all three were the case. He was smart — but not always smart enough to disguise that he was self-serving. He sometimes absorbed others' ideas into his own programming procedures

Bill Stewart in 1957. Stewart at one time had the title of National Program Director for the Storz stations. Because he changed jobs quite frequently, he held similar titles with other station group owners, some of whom competed against Storz. He was smart, creative — and not always loyal. Although he contributed to it, Bill Stewart did not invent the Top 40 radio format.

and failed to give credit where it was due. He is often thought of as one of the creators of the Top 40 format, but he joined KOWH two-and-a-half years after Top 40 was well established in New Orleans on WTIX, and thirty months after current hit music was the full-time mainstay on *all* the Storz stations. So Stewart was not a Top 40 pioneer, but his name on the pre-convention publicity, together with the mystique of sponsorship by the high-flying Storz stations, may have helped to boost attendance.

Todd Storz did not have the time to provide much oversight to the convention planning. He had been busy transferring KOWH staff to new assignments at WDGY and WQAM, finally closing the deal to purchase the higher powered WWEZ in New Orleans, and making arrangements to offer WTIX's original frequency to the local school board as an educational station — pending FCC approval. Todd Storz "had a full plate."

For this first conference event, full-page advertisements in major trade magazines promised attendees there would be "No Cost To You For Any Of The Functions ... including the talent-packed 'All Star Show' which follows immediately ... underwritten by America's leading record companies. Thus, your expenditures will be limited to transportation and hotel accommodations." Clip-out registration forms were to be mailed to Bill Stewart's attention in Omaha.

Stewart framed the initial Kansas City DJ convention in the language of a seminar offering proprietary information from high-priced media experts, radio-group executives, and major-market air talent. The 1958 "All Star Show" in the evening featured musical turns by Tony Bennett, Peggy Lee, The Four Lads, Lavern Baker, Don Rondo, Andy Williams, The Crew Cuts, and many others.

It turned out to be a hail-fellow-well-met convocation of radio voices competing for the attention of program managers who might hire them. They convened to hear Sidney Roslow of Pulse, a national ratings service; Frank Stisser, from C.E. Hooper, Inc., another ratings service; Gordon McLendon of the McLendon Stations; Harold Krelstein of the Plough Stations; Arthur McCoy of John Blair, a national station sales representative; and Adam Young of Adam Young, Inc., another national sales representative.

A Saturday morning breakfast opened the event with welcoming remarks by Todd Storz, followed by Kansas City mayor H. Roe Bartle, whose towering frame and big voice helped to set the tone that Kansas City was the home of WHB, the big flagship station in the middle of the nation that had been so instrumental in creating the Top 40 radio movement.

Martin Block, then a WABC New York disc jockey, reminded the audience it was he who originated disc jockey techniques twenty-five years earlier with his *Make Believe Ballroom* program on WNEW — although Block really borrowed both the concept and the name from Los Angeles radio personality Al Jarvis.

It would be a full day on Saturday, March 8, for panel discussions on such topics as:

> What the DeeJay Can Do for the Advertisers at the Local Level
> Increasing Income and Prestige through Related Outside Activities
> Are Today's Radio Ratings Services Obsolete?
> How to Run Profitable and Successful Record Hops
> The Ingredients for Today's "Formula Radio"
> Is Rock and Roll a Bad Influence on Teenagers?
> The Record Artist's Obligation to the Nation's Youth
> What the Time-Buyer Looks for in Buying a Market, and
> The Program Director: Friend or Foe?

Roundtable groups and panelists invited the participation of everyone in attendance. Pick a question of interest, and chances were good an expert on the subject could tackle it. Announcers, news voices, programming, production, promotion people, and sales-minded management types were looking for (or providing) answers. (Others, temporarily bored with it all, might seek out a prostitute strutting her stuff at the Muehlebach.)

The *Kansas City Times* reported the hometown event thusly:

> The successful disc jockey established himself as a personality to his listeners, a roundtable of record spinners agreed today ... if a listener [merely] wants to hear records he can turn to any radio station. WHB disc jockey Eddie Clark said, "radio men lead the listening audience, not the jukeboxes." Don McLeod of WJBK in Detroit put it this way: "You wouldn't put your money in a jukebox to play a song you hadn't heard, would you?" Panelists agreed that knowing when to stop talking is "the most important thing to remember. Ideal timing is 15 seconds for introducing a record, and about a minute between selections."

George W. Armstrong, newly named Storz Executive Vice President and WHB General Manager, had the rapt attention of everyone as he covered the topic "What I Look For in a Prospective DJ." He listed the qualities of "believability, sincerity, microphone integrity, and the ability to accept responsibility."

But not all the speeches painted a rosy and responsible portrait of the modern disc jockey. Mitch Miller, then head of Columbia Records' popular music division, charged the Top 40 purveyors in attendance as having "abdicated your programming to the corner record shop; to the eight-to-fourteen-year-olds; to the pre-shave crowd that makes up 12 percent of the country's population, and zero-percent of its buying population — once you eliminate ponytail ribbons, popsicles, and peanut brittle," reported the *Kansas City Times*. "At best," Miller said, "the Top 40 presents a philosophical problem on a par with 'which came first, the chicken or the egg?' Does the demand [for a certain record] come because you play it first, or do kids demand it because they find it on the Top 40?"

Top 40 broadcasters in the audience may have sensed that Mitch Miller wasn't fully aware that many of them were piling up huge ratings and raking in fat revenues, due in part to their playing the very music Miller was complaining about — a criticism aimed at the coming floodtide of post–World War II pre-teen and teenage consumers. But by the early 1960s, Top 40 operators would be scrambling to attract their share of a coming $500 million windfall in advertising aimed at the "youth market"—consumers of soft drinks, trendy clothing, health and beauty aids, shoes, pizza and other fast foods, movie passes, records, jewelry, outdoor recreation, and the like. In one of many Top 40 advertising success stories, Ford Motor Company chose the nation's leading Top 40 stations to help introduce — in the spring rather than the fall — what became its landmark "1964 ½" Mustang. Top 40s teenage listeners were morphing into eighteen-to-twenty-four-year-olds, and would grow into eighteen-to-thirty-four-year-olds, or "influencibles," as Storz would call them. Media and marketing moguls would later refer to them as young, upwardly-mobiles, or "yuppies."

To enthusiastic applause, Armstrong promised a second disc jockey and programming convention to be held in March 1959 in Kansas City. That plan was later revised when Stewart announced that the second convention would be shifted to late May, and would be held at the Miami Beach Americana Hotel near Storz's WQAM — the host station. Stewart's intention was to become Storz's right-hand man. His 1958 Kansas City post-convention follow-up and his 1959 aggressive pre-convention publicity and promotion served his own ambitions — or so he thought.

Bud Connell is one of a number of people in the radio business who feel that Stewart's contributions to Top 40 have been inflated. Connell's initial opinion of him had been formed in 1956, when Stewart took his time in arranging for Connell to join the KOWH air staff— almost six months after Todd Storz had made the initial offer. Says Connell of his first day at KOWH:

> Stewart gave me an unfriendly hello, spent approximately an hour with me on the morning procedures, and (this is the surprising part) I never saw him again until an early 1958 cocktail party in New Orleans when he was still Storz National program director, and I had become the new Program Director at WNOE. Bottom line, when I was in Omaha, I never saw a memo from Stewart, or any kind of a policy statement, let alone a programming operations guide.

Bud Connell may not have received any coaching from Stewart because Connell already had an excellent grasp of the essential elements of successful Top 40 programming. For less experienced air personalities, Stewart had developed a "cram course" in Storz programming procedures. When Dave MacFarland interviewed Stewart in 1971, he obtained a photocopy of two "Storz Station Operation" checklists that had been written in January of 1957 as a sort of "exam" by two men seeking to be hired by Stewart as KOWH air personalities. One was signed by the writer — Mel Leeds — while the other bore only the initials "dg." These "checklists" are the kind of nuts-and-bolts detail that the Storz Stations tried very hard to keep "in-house" and thus unavailable to competitors. Whereas the two Storz programming conventions which Stewart helmed were *about* programming, detailed explanations of station formats such as those below were attempts to establish the patterns and procedures *of* Storz programming.

STORZ STATION OPERATION PROGRAMMING:
Mel Leeds

Pace on air should be brisk. No chatter longer than 60 seconds between records.

Never segue from music to music. Can segue from ET ["Electrical Transcription," an older name for acetate disc recordings, to be explained later] to music, providing announcer talks over record to give artist and title.

Check for: dead air-key [switch] noises — background noise — conversation to unknown persons in studio — excessive chatter.

Re-cut any faulty ET's — and replace scratchy records.

Never telegraph [the] record before commercial. It is permissible to do so only before the news on show longer than one hour — then after beep, play record telegraphed. Use up-temp song in this position — preferably one in the Top Ten category — never use the #1 song.

Always close show with instrumental, in the event of fade.

#1 song to be played every hour on all shows.

Pick Hit of the Week to be played once during a show.

Insert Time and Temperature between records, preceded by call letters.

Insert Time and Temperature to split double spots.

Secret Word given on all shows. Must be given straight. Do not embellish. On echo.

Records or ET's never to be up-cut [cut off early].

Always hit the beep when giving time on the hour.

Frequently mention position of record on Top Forty list.

Never play or discuss flip side ["B" side] of record unless on the list, or chosen as an Extra.

Check log every day to see that all promos are listed, such as Weather Jingles — Station Breaks — Station Promos — Dr. Krick [syndicated weather forecaster] — Sound-Off — News Promos — and other inserts that you may include.

Announcer to sign log after going off air — for each show.

Monitor station at hours when least expected — 3:30 A.M., etc. Can tape shows at home to discuss mistakes on air.

Hold weekly meetings. Conduct with diplomacy.

Check with Music Librarian to see that Top Forty List is checked with local resources and then check with Omaha for coming week.

No show will ever accept requests.

Never apologize on air for anything done or undone.

#2 record and #3 record played at discretion of announcer.

Always refer to News Center — not News Room

Bulletins to be aired after music has started — so dramatic effect can be had by breaking into record — bring up record after bulletin.

Approximately 14 items to be used on newscast, in this order: Local/Regional — followed by Around the World/Around the Nation.

Announcer doing the news to read items before airtime to familiarize himself with contents and to shorten items when necessary.

Manufacture "Sound-Offs" if those sent in mail not useable.

Change news format every 4–5 weeks. Change other promos and station break tags every 2–3 weeks.

Send changes to Bill Stewart a week before airing — like gimmicks — ideas — promos — etc.

If announcer on competition station doing good job and is threat to you — get him if you can.

Plug "America's # 1 Independent Station."

To win Pick Hit of the Week record — explain on each show.

Can play 2 old pick hits plus 3 local extras other than the 7 extras sent from Omaha each week.

When adding new tags — promos — and gimmicks to shows — check with traffic and inform them of changes.

No shows to be taped, unless it's day off of Announcer. Otherwise *verboten*.

Keep in close contact with Announcers — Music Librarian — Traffic — Station Manager — Home Office, and Paymaster

Ask questions of Supervisors at local station or Home Office.

Be on time — keep your big mouth shut — and work like a bastard ... and you robbed me of my notes if anything went by unnoticed.

SELL THE SOUND — AND KOWH — OR CALL LETTERS

The second candidate (known only by the initials "dg") wrote his answers more like an operations manual than as a checklist. His answers were more detailed, and over two days, he covered the same number of topics (39) as Mel Leeds.

LEARNED TO DATE: KOWH 1/7/57
                                                      dg

STORZ PROGRAMMING provides: A bright, briskly-paced show. Top tunes at all times. Tunes selected from the Top Forty (with record shops, distributors and coin operators polled locally) plus ten "extras" plus "Pick Hit of the Week" plus two old "Pick Hits." Currently, five rock 'n' roll have been added. A "Hit of

Yesteryear" is planned. An Album of the Month with one tune used per day has just been added.

TUNE SELECTION: Top Forty compiled each Wednesday in KOWH, Omaha by correlating information mailed (Air Mail Special Delivery) each Monday from Storz Stations. Stations submit their Top Forty plus 3 local favorites. Omaha returns master list (minus rating numbers) of Top 40 plus 7 "Extras." If 3 local favorites are missing, they are to be inserted at spots they deserve.

AIRING TUNES: Only tunes on list to be played. Only listed version to be played. Announcers instructed not to apologize for not "having" or being "allowed" to play other records. Announcers not to discuss "flip" sides. Requests are not accepted (lone exception, Sandy Jackson Show on KOWH). Frequent reference to tune's position on (Call Letters) Top 40 to be made as tunes are used. Special intros (E.T.) are provided for No. 1, 2, and 3 tunes plus "Pick Hit of the Week." These are to be used every time corresponding tunes are aired. *Number 1 tune must be aired once every hour! Same for "Pick Hit of the Week." Number two tune should be aired once every show.*

AIRING: Shows should open with bright, up-tempo tune. Close with an instrumental (so fade is possible if necessary). SEGUE FROM MUSIC TO MUSIC IS FORBIDDEN. *Announcer may segue from E.T. to music only if he comes in over for intro.* CALL LETTERS, TIME AND TEMPERATURE TO BE GIVEN BETWEEN EACH RECORD. In case of double spots, call letters should be given as, "KOWH time—"; E.T.; temperature as, "It's now 00 degrees," and E.T.—*No more than sixty seconds of chatter between records.*

Announcers should telegraph the return of their show (*following news at five minutes before the hour every hour*). Can use: "We now invite you to take five minutes out for the latest KOWH news—then we'll be back with the number 00 tune from this week's KOWH Top Forty—(artist's name) and (tune title). Announcers to bang music directly following the beep (on show come-back). Wait until after first tune to "welcome back" the audience. Announcer should be wearing headset (cans) and be ready to come right in on top of E.T. or record. There should be no pause. NEVER DEAD AIR. *Announcer must avoid up-cutting at all times.*

Announcer to give weather forecast (The KOWH weatherman says—) every four or five records. Complete weather forecast is given with (following) the news every hour. Storz Stations now featuring the private weather forecasts of Dr. Irving P. Krick—providing exclusive "two" outlooks on the weather ... official Government reports and Krick. Live and E.T. promo spots provided. Are to be inserted with Krick plugged every hour.

NEWS SHOWS: Special news sound effects are included in News Formats. These to be changed every 4 or 5 weeks. Complete change made, including lead-in and etc.

NEWS DELIVERY: Announcers to be on hand half-hour before doing the news. News to be pre-read. News room always referred to as NEWS CENTER. Announcers instructed to keep items short. To include 12 to 14 items per newscast. Sequence—LOCAL ITEMS FIRST—REGIONAL—J. MILLER E.T.—NATIONAL AND INTERNATIONAL. "SOUND OFF" (mail or manufactured) included currently in the newscast. Two per day provided and they are to be used in rotation. Four per day rotated at 19-hour stations. News sound effect used in close. Next newscast is always telegraphed. *DATELINE* OF EACH NEWS ITEM TO BE ECHOED. Button on right side of news desk to be held down for full word ... a split-second pause then into item. News machine sound under newscast.

ANNOUNCERS: Instructed to direct talk to listener. No window-talk (conversation to persons outside the studio). They are not to discuss past history of artist or tune at length. Avoid fluffs at all times. If made, better to move on than stumble with correction (in case of between-record chatter).

JOHN MILLER FEATURE: Arrives on tape and is cut apart and placed on E.T.s at Kansas City. Care to be taken to use cut called for on log. Each opens with key-sound. Closes same way. Usually three, four items to a cut.

WEATHER: Currently non-key [sic] sound used behind all weather forecasts.

ANNOUNCERS: Must avoid studio noises, paper shuffling and etc. Must use care in avoiding key-click [switch noise] coming on.

PROMOTION: Promotion currently in use includes: station breaks and tags (changes every 2 or 3 weeks). "SOUND OFF" promos (changed every 2 or 3 weeks). A special promo kept cooking at all times (KOWH currently running a Dog Whistle promo). PROMOTIONS SHOULD CREATE CONVERSATION. They need to be "big" but must have the "talk-starting" ingredient. STORZ PROMOTION EMPHASIS ALWAYS DIRECTED TO "ON AIR" APPROACH. No interest in outside promotion. STATION PAYS — PROMOTION TIE-IN PAYS. NO DEALS — TIME FOR SERVICE, PLUGS & ETC. Policy on recognized Charity appeals: tie-in with all, to some extent. Time given as schedule permits.

PROMOTION: Current list also includes: Intros (E.T.) for #1, #2 and #3 tunes plus "Pick Hit of the Week." These to be added-to frequently and rotated. News promos — plugging KOWH news every hour, 5 mins. before the hour, plus emphasis on LOCAL & REGIONAL news. Bid for contributions of Best News Tip of the Week with $25 or $10 awarded — have been used recently. Weather promos — (2 types) plugging dual weather reports of Government and Krick; spots devoted to Dr. Krick.

PRODUCTION SPOTS: Production sessions held each morning (as required) at 10:30. Production chief in charge (at KOWH it's Grahame Richards). All copy for E.T.s placed in special bins by Continuity for pick-up by Grahame. He selects announcers needed and gets necessary sound effects together. As ready, promotion E.T.s would be cut following regular Production session. Copy changes to be made only after checking with agency or salesmen (on commercial spots).

DUTIES OF A PROGRAM DIRECTOR OF A STORZ STATION: Monitor station. This to be done at odd hours. Compliments as well as missiles to be hurled. Attention to new men on new shifts important — when clock is set and show caught ... CALL STATION — LET ANNOUNCER KNOW YOU ARE TUNED. Next time his show is check, drop over to a later hour. KEEP THEM GUESS AS TO WHEN YOU'LL BE LISTENING.

KEEP EAR PEELED FOR GOOD ANNOUNCER. Check with Station Manager. If interested, try and set deal.

CHECK ALL PROMOTION WITH STATION MANAGER — until instructed to do otherwise. Once a relationship has been established — showing copy prior to airing will prove sufficient.

SEND PROMOTION IDEAS to Bill Stewart, KOWH, Omaha at least a week in advance (competitive stations monitor STORZ STATIONS and it's possible a promotion would be picked up and aired by another station in a STORZ city before STORZ STATION could introduce it).

SEND CARBONS OF PROMOS to Bill Stewart, KOWH, Omaha as they go on the air. Keep in contact. Exchange ideas.

MONDAY — Compile Top Forty list — add 3 local extras and send via Air Mail Special Delivery (call to tabulate list, typing and mailing now handled by librarian at station — *see that it's done*). Set up for obtaining records already established at station — *follow it*.

TAPE WEAK SISTER SHOWS AT HOME.

ANNOUNCER MEETING TO BE CALLED WEEKLY — under direction of Station Manager. Opportunity for each announcer to air gripes, problems, & suggestions.

PROMOTION MONEY AVAILABLE — originate ideas, check STORZ ACCEPTANCE — if it's granted, wheels will turn.

GIVE-AWAYS: Only give-aways currently in use — 5 copies of "Pick Hit of the Week" awarded each day ... plus brand new "Album of the Month" promotion (1 per day — PROMOTION REPLIES TO BE LIMITED TO POST CARDS WITH NAME AND ADDRESS (for ease in selecting winners). PRESENTLY GIVE-AWAYS NOT BEING EMPHASIZED.

STORZ STATIONS trade paper advertising budget for 1957 totals $100,000. Effective display use can be made of these tear sheets.

THE GROUP SPIRIT OF STORZ STATIONS BRIMS WITH ENTHUSIASM.

FIVE STORZ STATIONS NOW OPERATING. SEVEN STORZ STATIONS ANTICIPATED.

POLICY IS TO HIRE THE BEST AND THEN HAVE EFFECTIVE OPERATION WITH MINIMUM NUMBER OF EMPLOYEES.

NEWS: *Bulletins — always held and put on (with special E. T. intro) as "break-in" during Music*. Transitional words *not* to be used: "Again, Washington–"

LOG: Check log daily to see that promos are scheduled — check also for announcers signatures on Original copy. *MUST BE SIGNED.*

SECRET WORD: "Secret Word" is a daily gimmick. To be obtained in advance from Sales Dept. Typed and names given month in advance to Continuity. "Secret Word" goes on upper right-hand corner of log — *to be given without embellishments* i.e. "Today's Secret Word — Tom Edison."

HOLIDAYS: Keep on top of Holidays. Prepare special spots a needed. Schedule them. See that they are used.

FAULTY E. Ts: Listen for faulty E. Ts. Go to engineer and have new cut made.

SUB-STANDARD RECORDS: Listen for poor quality discs. Replace.

(Error:) "Pick Hit of the Week" is *scheduled once per show* — not per hour. Correct original mention.

TAPING SHOWS: Announcer are permitted to tape shows *only on day off* — or in cases of extreme emergency.

ADDITIONAL INFO REGARDING "PICK HIT": Bids for mail made on all shows, as it is aired ... post cards to be addressed to STAR — STATION. AWARDS (5 per day). MADE ONLY ON "TOP FORTY SHOW" airing at 4 P.M. on all stations.

SELECTION OF "PICK HIT" WINNERS: (similar for all). Cards are dumped in waste basked (by librarian) and five pulled. To these is sent a processed letter, to wit:

(Date)

This entitles the bearer to pick up one free copy of the KOWH "Pick Hit" of last week, "Wringle Wrangle" by Fess Parker. The record may be obtained by presenting this letter to NAME AND ADDRESS OF SUPPLY *CONTACT*.

Thank you for your interest in KOWH, and we hope that you will continue to enjoy our (daily) programs which are designed with you, the listener, in mind.

Very truly yours,
Bill Stewart
PROGRAM DIRECTOR [*End of dg's exam*]

The preceding exhaustively-thorough listing of Storz stations' program elements and production procedures should lead the reader to the following conclusion: At Storz stations, producing and presenting the Top 40 show required far more than merely handing a guy a stack of records to play, and a sheaf of commercials to read. There were complicated protocols and a myriad of practical procedures to be learned, mastered, and constantly retrofitted if both the air personality and the radio station were going to prevail in each market. Following those procedures required energy, enthusiasm, stamina, talent, and brains. But in the final analysis, Stewart's "Station Operation Checklists" were only a safety net — a measure of assurance that a newly arrived Storz announcer could use the procedures to generate a show that at least was consistent with the format and met audience expectations. The higher goal was to become a true "air personality" who so engaged his audience that he might "violate" the format and still post strong ratings. When all of the desirable format attributes were operating, there was a good chance that genuine entertainment would come burbling out of your radio whenever you tuned to a Storz station.

Speaking generally about Storz's program formats, Bud Connell insisted that "there was never a company-wide format. There was an *attitude*— when Todd's overall feel, transmitted through Bud Armstrong, gave the staff the sense that they could accomplish anything." Connell confirms that *not* reducing Storz programming procedures to a written form was purposeful:

> It was general Storz policy (unwritten!) not to have details in book form or otherwise that could easily be copied or pirated by competitors, by moles on the staff, or carried away by people leaving Storz employ. I recall light discussions with Bud Armstrong and [program director] Grahame Richards affirming that no "Programming Policy books" should be created, and if they existed, they would

make us vulnerable. I always meted out my programming policies in bits and pieces by memo, sometimes as many as six or seven a week, each memo usually covering only one point or one promotion. All the rest of the programming policies — new ones and reaffirmations — were transferred to my on-air staff during weekly personnel meetings. I believe, but I am not sure, that this was a kind of unwritten understanding throughout the Storz chain especially when Todd was alive, and for a while after his death.

The fact that Storz programming policies usually were *not* committed to paper caused other broadcasters to glean as much information as they could from people who had worked for Storz. An example of such "competitive information seeking" is evident in excerpts taken from an exchange of letters between Gordon McLendon's National Program Director Don Keyes, and Kent Burkhart, who had just left Storz's WQAM. The original correspondence, discovered by Dave MacFarland in the McLendon stations' archive in Dallas, is dated March 9, 1959. In it, Kent Burkart is responding to questions about Storz procedures posed to him by Don Keyes.

1. *What policy exists on how much talking a disc jockey is allowed to do between records? What is the theory behind this policy?*

A: The "talking" policy you refer to has never been in written form; however, it is governed mostly by the various markets and their radio environments plus a general policy of "don't talk unless you have something useful to talk about." The quoted rule is pounded into the head of each disc jockey before he begins work at the station, and at every available moment, or any *needed* moment. However, in some cases — for example, Kansas City — it is permissible for a personality to become cleverly chummy with his audience, but not in great length. Miami, Minneapolis, and Kansas City all observe this policy. New Orleans varies, as later words will indicate. Theory: useless words are truly useless!

2. *How long is a record kept on the air?*

A: A record is kept on the air for a period of not more than 16 weeks, or four months. A one week listing on the playlist is the minimum airtime devoted to a recording. However, as I will explain about the familiar or "recognizable" tunes later, I shall leave the preceding statement as it is.

3. *What is the policy on frequency of use of call letters?*

A: Actually, the policy on call letter mentions is up to the local management and program director. Obviously, both know the value of such frequent use of the call letters, and I believe I would be safe in assuming that at least one mention between records would be the minimum. In Miami, I required at least three mentions between every record during Hooper week, and at least two mentions during the other three weeks of the month. These were live announcer mentions, and did not include any ET'd [recorded] promos, singing breaks, etc. In brief, hot and heavy use. All of the station observe the same or similar mentioned ideas.

6. *When Todd buys a station, what does he generally look for? That is, power, frequency, etc.?*

A: His main interest has been low frequency with five or 50,000 watts. That's why the KOMA buy shocked the entire company. He doesn't seem to be too concerned about competition in the town. He feels he has the horses anyway and with a good, established, old-time, low-frequency operation, he can grab the ears of many new listeners. (That is the reason for the switch [to a higher-powered station] in New Orleans, obviously.) He converts into a fast-paced operation while still holding on to the old call letters as a prestige credit. This has been very successful! (The KOMA move is still not clear and I have not heard or seen any stock movement or offer from anyone.)

9. *You and I discussed programming "recognizable" tunes. Can you explain this idea more fully?*

A. Well ... in general the idea is to play tunes that are currently popular or have been popular, not those necessarily that will be popular. After all, we're trying to entertain the people listening, not force them into something new that they may not and often do no like! KLIF follows this idea, maybe not knowing it, and so does KILT. However, KTSA sounds at times like the Hit Parade is a thing of the past. In truth, since you came down and goosed 'em, they sound a hell of a lot better. (You should do that more often.) Incidentally, this is a strong Storz belief, and I can safely say that only about five to ten new records a week are new ones. Those are selected from the obviously up-and-coming records listed in *Billboard*'s "Coming Up Strong" section.

10. *Is there a commercial limit set within a given hour of airtime?*

A: Yes. Eight commercials, minute length, first half of hour; seven commercials, minute length second half of hour, and six twenty-second announcements, two of which are scheduled on the half hour break, two of which are on the break before the news, two of which follow the news. All are separated by a singing station break and perhaps a time or temperature check.

11. *How many salesmen are there at each station?*

A: This varies with the manager's tastes. Storz likes many. In Miami we carried six to eight at all times. In Kansas City I'm sure it's about six. In Minneapolis the same, and in New Orleans about three or four I would guess.

12. *Does the Top 40 show in the afternoon run on all stations? Does the disc jockey start out playing # 40 and go through to #1 each day?*

A: At the time of my leaving [WQAM], the Top 40 was being run from either three to six, or four to seven at all the stations. This has been the most popular radio program, in whole, for the Storz stations. The idea was stolen from Bob Howard's Top Twenty in New Orleans years ago. Let me point out that the format of each Top 40 program will vary with the individual stations. Experimentation is used to find out whether 40 through 1 or a rotating record show will better identify a larger audience. In Miami, Monday-Wednesday-Friday, it's 1 through 40, and Tuesday-Thursday-Saturday and Sunday, it's 40 down to 1. Kansas City, I believe is still rotating the records in any order. The same applies for the Minneapolis station; again, I'm not too sure about the New Orleans station. When I was in Omaha, I actually ran the first 40-down-to-1 every day type of operation. It was a success there. We had 50 and 60 percent of the audience; however, we found a drop in Kansas City, etc. Depends on the market.

14. *Is there any pattern within the hour on the use of time, temperature, weather?*

A: I required temperature and time between every record, weather forecast every 15 minutes, plus twice on the hour; however this is up to the local Program Director. There is no set pattern as such.

15. *How important does Todd consider sports scores to be? News, especially local?*

A: Sports scores have really never been stressed except in Kansas City (major league). They are generally carried on the :25 mark, and if a big score comes in, it will be given immediately. But, in general, no promotion is given. The Storz stations try to build news reputation by spending a *little* money and by promoting the news director alone … (always the A.M. news man). Mobile units are too expensive according to Omaha so most of the units have been done away with. The news director tries to get close to the desk sergeants by sending them a fifth every now and then … but no real money is spent on news at all. Matter of fact, the recent "economy wave" kicked out one of the news services plus the special news telephone ("hot line news tips") plus the weather and Western Union machines.

18. *How do you explain the difference in the sound of WTIX and other Storz stations? Why is it so much tighter and faster and more formalized?*

A: This is very simple.… They are instructed to say nothing but the time and temperature, read the plug board, comment on another disc jockey, etc. They are told to move quickly — fast pace. Since there isn't much chatter anyway, this produces a quick, tight, fast operation. That sounds ridiculous, but has so far been doing OK. Fred [Berthelson, WTIX Manager] is a big one for screaming and hard pitch on the air also. Now you can see.

The correspondence between Keyes and Burkhart shows that even well-run group owners such as the McLendon stations were eager to learn particular details of the Storz "success formula." Following the initial Pop Music Disc Jockey Convention and Radio Programming Seminar, fully emulating Storz Top 40 was the most a broadcaster could hope to do.

• SEVEN •

Programming Conventions II: Tarnishing the Top 40, and Touting "Talk"

The 1958 Kansas City Pop Music Disc Jockey Convention and Radio Programming Seminar had been a great success, so the Miami Beach follow-up was expected be a "whopper"—and it was, although not for the best reasons. The 1959 convention attracted many more attendees—an estimated 2,500 people—but it also became the springboard for brewing allegations of "payola"—the practice by some disc jockeys of requiring "pay for play" to get a new record aired. WQAM's Jack Sandler—the general manager of the second convention's host station—was subpoenaed by U.S. House of Representatives payola investigators. A similar subpoena was served on Miami Beach's posh Americana Hotel for delivery of its records covering the convention, at which disc jockeys were generously wined and dined. The *Miami Herald* (which had sold WQAM to Storz just three years earlier) distanced itself from the station with the soon-famous, scathing headline "Booze, Broads and Bribes" in characterizing the event. Unfortunately, all three words were factual.

The "bribes" accusation was taken up in Washington by the House Legislative Oversight Committee, which wanted to know the relationship between the Storz stations and eighteen record companies that picked up a $117,000 tab (about $904,394 in 2012) at the Americana Hotel. Eleven years later, a retrospective article in *Broadcasting* put the best possible spin on what the Storz Stations had tried to do: "Todd Storz and his station group were sponsors of the first annual disc jockey convention in Kansas City. (The second held the next year in Miami Beach, was the last. Record company entertaining got out of hand.)"

According to Storz general manager Deane Johnson, "There is an 'urban legend' that at the Miami Beach DJ convention, when Todd was getting pretty uptight over what was going on, and was apparently on Bill Stewart's case, Stewart said, 'For two cents, I'd just quit.' Todd is reported to have reached in his pocket and handed him two cents. I have no idea if there is any basis in fact for that story."

Because of the improprieties and the accompanying negative publicity surrounding the second convention, Bill Stewart *was* terminated by Storz, even though (according to Bud Connell) unlike his father, Todd was not someone who fired a man easily:

> I am not aware that he ever fired anyone [else]. The philosophy that seemed to permeate the company was to hire the right people and give them the space and support to do the job. In my opinion, and considering Stewart's rocky past with Storz and other employees, Todd gave him every benefit of the doubt ... but his [Stewart's] string simply ran out.

After being fired by Storz in 1959, Bill Stewart went to work at Don Burden's KOIL in Omaha, which had been KOWH's chief competitor when KOWH was still owned by Storz. In 1961, Stewart moved to Dallas where he was employed for a time by the PAMS jingle company (which had exclusive contracts with all of the Storz Stations), and then was a consultant for the McLendon stations. After the dust from the second Disc Jockey Convention had settled sufficiently, he returned to Storz as national promotions manager (although Stewart's resume says "Program Director") between April 1964 and November 1966. In that job, he was based at WHB in Kansas City. Stewart had the misfortune to begin work there on April 13, 1964. It was the same day that Todd Storz unexpectedly died in Miami, as will be detailed in Chapter 12.

Deane Johnson remembers that during the 1964–1966 period, "Stewart 'meddled' with WTIX, and station manager Fred Berthelson called Bud Armstrong to protest his presence — and the fact that I (as program director) refused to deal with him. Armstrong's response was rather revealing when he said, 'I figured I'd have to get involved in these kinds of situations. I'm not going to lose a good program director over Bill Stewart.'" Stewart resigned the promotions manager position and left WHB in October, 1966, returning to Dallas where he again worked for Gordon McLendon's stations. (Two decades later, on December 6, 1985, the *Dallas Morning News* reported Stewart's death at the age of 58, saying that he was "instrumental in introducing Top 40 radio programming in the early '50s. During the 1960s and '70s, he acted as creative consultant for broadcast groups throughout the country, winding up in the real estate business as a broker." While no cause of death was given in the newspaper's obituary, radio gossip had it that Stewart had taken his own life.)

In the late 1950s, the exploding nationwide popularity of the Top 40 music format prompted the music industry to produce more records that would especially appeal to those hit-music listeners. Because of the growing glut of new records vying for attention, record promoters regularly visited stations or lobbied disc jockeys privately to get their records played. In 1959, "lobbying" was legal, and so was being "paid for play." The nickname for being paid for radio airplay of a certain record was "payola." The reputations of pioneering radio disc jockey Alan Freed, and of Dick Clark, host of ABC television's *American Bandstand*, were both tainted by payola scandals. In 1959, Freed was fired from WABC, New York, and in May 1960, he was indicted for accepting a $2,500 "token of gratitude" (about $19,228 in 2012). Freed's career went into a tailspin from which he never recovered. He did not work at another major station until his premature death in 1965.

For the majority of the U.S. radio industry, the solution to the payola problem was to take the decision — about which records to play — out of the hands of disc jockeys, and place it with upper radio station management, who became responsible for creating the station's playlists. The presumption was that management-level program directors would have more concern for the welfare of the station's FCC broadcast license than could be expected of disc jockeys.

In 1960, following months of hearings and sensationalized news reports, a U.S. House subcommittee report found that "207 disc-jockeys in 42 cities had received over $260,000 (about $1,975,816 in 2012) in payments ('payola') to play records on the air." In New York, a grand jury indicted eleven disc jockeys for commercial bribery. Further, twenty-three record companies admitted making payments. Soon, an anti-payola statute was passed that carried penalties of up to a year in prison and a fine of $10,000 (about $76,913 in 2012).

In the wake of the payola scandal and the resulting Congressional oversight, an announcement crafted by Herbert S. Dolgoff, general counsel for the Storz Broadcasting Company, was prepared for broadcast on each Storz station, as follows: "Certain records heard, or given away on [Storz station call letters], were provided in consideration of cooperation by various record artists, manufacturers, and distributors." Wording to the same effect was used by other Top 40 stations. It wasn't an outright admission of receiving "pay to play," but it certainly wasn't a denial either. Similar phrasing is included in the "credits" of a surprising number of modern-day radio and television programs — another legacy of the Top 40 era.

It needs to be emphasized that the payola scandals were mostly concerned with the teenage-appeal music being played by Top 40 stations. The fact that a relatively large slice of teens' "disposable income" was being spent on hit records was of less importance than the fact that some of the members of Congress running the hearings abhorred the content and style of certain records and/or the performances of artists who were deemed crude or lewd. The discovery that some disc jockeys personally profited from playing such "filth" was the last straw. Lost in the congressional grandstanding and demagoguery was the fact that plenty of "good" records from respected musicians made their way onto Top 40 playlists without money changing hands. For example, there was no evidence of a representative from RCA Victor Records paying a bribe to get a Perry Como song like "Catch a Falling Star" played — and yet "legacy" artists *were* played on Top 40 stations. In fact, Top 40 playlists were far more eclectic than most of the format's critics would admit. Teens were a dependable, novelty-driven audience, but their spending power was limited. Adults with bank accounts were the listeners whom radio advertisers wanted to reach, and Storz stations' music programming was designed to deliver them.

Marc Fisher, in his book *Something in the Air*, says

> Those who lived through the payola scandal came to see the purging of rock radio as the older generation's desperate effort to hold on to what they knew, to their ideas of how parents and children should relate to one another, to their concept of race in America, to their sense of respect and propriety. "The whole thing was about resentment of the music," said Steve Labunski, who ran Todd Storz's Top 40 station in Minneapolis. "If they were paying people to put Glenn Miller on the radio, they would never have stepped in. The payola scandal let people get off their chests their resentment of youth and their music, and blacks" (p. 89).

Fisher also offered details of how a Federal Trade Commission investigation of hundreds of deejays led to a substantial number of them deciding to quit before they could be fired. His conclusion is heavy with irony:

> But there would be no wave of prosecutions. And payola would never really go away; it merely changed direction. Now it was music directors and station managers, rather than deejays, who made deals with record companies and their distributors. The same free records poured into stations. The same influence was peddled. But stations slipped disclaimers on the air at odd hours of the night: "Certain records heard or given away on KXOK were provided in consideration of cooperation by various record artists, manufacturers and distributors." The payola scandal's main impact was to cut deejays out of decision making about the music that got on their air (P. 91).

With the record industry/radio airplay brouhaha keeping Top 40 stations under a cloud of suspicion, Storz managers at some stations were glad to be able to point to non-musical programs which occupied their late-night hours as proof that they were taking their "public interest" responsibilities seriously.

It is not well-known that Storz's Mid-Continent Broadcasting Company (later renamed the Storz Stations) were among the early adopters of what is now known as "telephone talk" programming. Barry Gray of WMCA, New York, is often credited with being the first to broadcast telephone interviews, beginning back in 1945. Alan Courtney began hosting a call-in program on WGBS, Miami, starting in 1948 — moving later to Storz's WQAM. However, the first iteration of non-music programming on a Storz station was not a telephone call-in show, but rather a "roving microphone" program on WHB called *Nite Club of the Air*, which began regularly scheduled broadcasts in October 1954. It originated from the coffee shop on the ground floor of the Pickwick Hotel at Tenth and McGee in downtown Kansas City, Missouri. WHB's studios and control room were on the eleventh floor. Hosted by disc jockey Wayne Stitt, the show featured popular music and roaming interviews among people who showed up in the coffee shop between 10 P.M. and 1 A.M. A revised version of the show, called *Nite Beat*, introduced a new kind of host whose job it was to stir up controversy and converse with local visitors, guests who joined the program by long distance telephone, and local radio listeners who called in their comments. All of the coffee shop chatter, plus local and long distance telephone conversations, were directed through a unique multi-path audio distribution amplifier that was not yet commercially available to broadcasters. The specialized unit, dubbed "Multi-Phone," was conceived and built by Dale Moudy, Storz's director of engineering. It allowed the show's host, guests in the coffee shop or the studio, and callers on the phone to hear each other and have their voices be "on the air" simultaneously.

Also part of the *Nite Beat* show's hardware was a seven-second audio delay, to reduce the risk of foul language getting on the air. In a normal tape recorder, the recording tape first passes by an erase "head" to delete anything previously recorded. The tape then moves on to the record head where the audio is laid onto the tape. Finally, it passes by the playback head. In this standard setup, the amount of delay between recording and playback is less than a second — not enough time to "dump" any offending words. Dale Moudy's audio delay unit employed a tape recording deck — mounted in a large equipment rack. The deck had its recording and playback heads reversed: the playback head was located *before* the erase and record heads. A 52-inch-long loop of recording tape was spliced together and routed through quarter-inch-wide guides mounted at key points around the perimeter of the audio rack. The signal from the audio console was recorded instantaneously at the recording head, but because it was the *last* of the three heads, it took about seven seconds for any naughty word to pass through the guides mounted all the way around the rack, reach the playback head and go on the air — plenty of time to "dump" the offending audio. It was necessary to replace the tape loop daily because after several hours of the same 52-inch loop of tape passing through the heads and guides every seven seconds, it was worn out.

Storz offered *Nite Beat* and nightime telephone talk programs on his stations as counterprogramming to the 10 P.M. (Central time) television newscasts—with their visual impact, local news film coverage, on-the-spot reporters, and movie-star handsome TV anchors. Storz's intention was to re-aggregate his stations' evening audience being lost to television news, but he may also have wanted to give adults a reason to tune his stations for stimulating talk rather than hit records. If nothing else, listeners to *Nite Beat* would likely turn off their radios without changing the tuning. They would wake up to Storz Top 40 programming the next morning.

In the early days of telephone talk, it was no small task to find a moderator who could research discussion subjects, recruit qualified guests, and work the radio audience as a well-informed program host. In Kansas City, Bud Armstrong found their man in Lee Vogel, whom Storz described as

> not only a very talented and civic-minded performer, but also an extremely well-informed person. He is an alumnus of Carnegie Institute of Technology, having majored in English and History. He has a long and varied background in commercial radio broadcasting, and ... is an ideal choice to conduct the unique nightly program.

Storz included that description in a 1957 study of a forum-type discussion program, which became part of the "programming ascertainments" section of WHB's 1959 license renewal application to the Federal Communications Commission. A list of the subjects actually covered by Vogel on *Nite Beat* were quite a bit less lofty, reading more like a grocery store tabloid:

> "Hypnotism"; "Does Indecent Dress Incite Sex Crimes?"; "U-F-O Study Club"; "American Students In Moscow"; "Nike Missile Installation Around Kansas City"; "Witchcraft"; "Why Women Should Not Get Married"; "Prosecution of Teen-Age Delinquents"; "The Communist Party"; "Gambling In Kansas City"; "Integration in Schools"; "The Danger in So-called Wonder Drugs"; "Profile on James Hoffa"; and hundreds of other topics.

The *Nite Beat* concept began to be copied in 1958 by other stations in the top 25 markets. Page 357 of Storz's 1959 license renewal application stated "*Nite Beat* has become so valuable a part of Kansas City community life that applicant has received scores of requests from radio stations all over the country seeking permission to emulate the procedure."

In Miami, WQAM's telephone-talk program was promoted under the name *The Alan Courtney Program* because Courtney had moved his show from WGBS to WQAM. Courtney enjoyed his recognition as "the number one rated night-time show in Miami," according to *Pulse*. In its 1959 license renewal application, Storz claimed Courtney's program to be "the highest rated public service show in the country."

To launch the show in the Minneapolis–St. Paul market, WDGY brought in Lee Vogel from Kansas City's WHB. Replacing Lee Vogel on WHB's *Nite Beat* was Tom Jacobson. Only in his mid-twenties, Jacobson had a powerful baritone delivery, sounding much like the TV anchors he was competing against. He would intone, "I am Tom Jacobson. Every telephone in Mid-America is your ticket to free speech the *American* way. WHB Multi-Phones are open tonight as a nation listens to you on ... *Nite Beat*!" After failing to show up for work one night, Jacobson's body was found at his home in eastern Jackson County, Missouri, with a bullet in his head. It was an apparent suicide. That night, WHB program director Don Loughnane replaced the absent Jacobson with this sizzling Storz-style opening delivered in a judicial tone of voice: "Tom Jacobson is dead!"—a sensational tabloid-style headline if there ever was one.

In Minneapolis–St. Paul, WDGY News Director Ralph Martin became *Nite Beat*'s host, loading his 10 P.M.-to-midnight show with major personality interviews, including big stars such as Pat Boone. Unlike the procedure at WHB, on WDGY all calls were taken live. There was no screening. There was no producer. Perhaps coincidentally, one night Martin was rushed to the hospital suffering from a nervous breakdown.

The big voice of WDGY's 21-year-old Jim Ramsburg took the *Nite Beat* reins for the rest of the year, in addition to his midday music show. "With Multi-Phone and the seven-second delay, all I had was an engineer; and I was on my own," said Ramsburg. His very first program topic introduced two guests—a Jew and a Palestinian speaking about the boiling conflict in the Mideast. Ramsburg's authoritative voice and mic-side discipline somehow kept things in control. WDGY Program Director Don Loughnane instructed the young talk-meister to "play the listeners like records." That may have meant that he should keep moving to the next call rapidly, since hit records in that era were often quite short. Station Manager Steve Labsunski, using his index finger in a circular motion, encouraged Ramsburg to "stir it up." WDGY's biggest fear, said Ramsburg, was "when the phones were not lighting up.'"

Telephone talk was not programmed at the two stations purchased later by Storz: KOMA in Oklahoma City and KXOK in St. Louis. But it is clear that the three Storz stations discussed in this section—WHB, WDGY and WQAM—helped to lay the foundation for contemporary telephone-talk radio, beginning in 1957. They created the fundamental procedures for effective audience participation in radio programming with the Multi-Phone audio distribution amplifier, the tape delay, and controlled program content — based on the precept that "every home is a studio, and every telephone is a microphone." The logistics necessary to allow that phrase to be true were daunting. But the objective was simple: To maintain or restore Storz station listeners (especially adults) who might otherwise tune in to the 10 P.M. television news — and to have the radio tuned to the Storz station's frequency when it was turned on again the next morning.

• EIGHT •

The Air War in Oklahoma City: KOMA vs. WKY

In 1955, Paramount Pictures released its Cold War thriller *Strategic Air Command*, starring James Stewart and June Allyson. The premier showing took place in SAC's hometown of Omaha with a gala party honoring SAC personnel and its commanding officers, including General Curtis E. LeMay. The party, with plenty of Storz Beer and other libations, was held at the home of Arthur C. Storz, Sr., in the Storz family mansion.

In the mid–1950s, thousands of eyes were looking skyward from observation posts across America for suspicious aircraft "in the wrong place" which might be Soviet bombers carrying atomic weapons. Those eyes belonged to the men and women of the Ground Observer Corps, an air force–activated civilian volunteer group whose "spotters" reported sightings to regional air force "filter centers" for evaluation and confirmation. Mostly what they saw were the high-flying contrails left behind by America's fleet of B-47 jet bombers — soon to be replaced by the larger B-52. The U.S. bomber fleet was capable of carrying the same nuclear weapons that Americans feared the Soviet Union would try to unleash on the United States. Nuclear armageddon was — at least to some — thinkable. So perhaps it is not surprising that in 1956 alone, 35 million prescriptions for the new "tranquilizer" drugs were written for anxious Americans.

It was "Cold War hysteria" that prompted the federal government to construct a bombproof studio at KOMA, the Oklahoma City station that was acquired by Storz in 1958 for $600,000 (about $4,670,014 in 2012). And it was the popular music and upbeat presentation of the Top 40 format on *two* strong Top 40 stations in that city which would keep a radio "air war" going well into the mid–1960s.

In 1961, when Deane Johnson sought to join the Storz Broadcasting

Eight • The Air War in Oklahoma City

Company[1] he felt he had a "pretty slim" resume to land a position with such a prestigious company. His previous experience had largely been in small markets in the upper Midwest. Johnson had programmed KWMT in Fort Dodge, Iowa — a Top-40-style station that had a powerful signal, albeit from a small market. As luck would have it, Johnson addressed his resume to Storz national program director Grahame Richards, who happened to listen to the Fort Dodge station and liked it, so Johnson's material got his immediate attention. Bud Armstrong checked Johnson's references. A couple of days later, Johnson was on a plane to Oklahoma City to meet KOMA general manager Jack Sampson, who hired him on the spot. The fact that Johnson also had earned a First Class radio engineer's license was a plus, since KOMA's high-powered transmitter required a First-Class engineer to monitor it at all times.

The transfer of KOMA to Storz Broadcasting had been held up for awhile, mired in the controversy at the FCC over whether the popular Storz Treasure Hunt contests were in the public interest. The sale was approved by the FCC in October 1958, with Storz taking control of KOMA a month later. Jack Sampson, who had been WHB's sales manager since 1956, was named general manager.

Sampson remembers several unusual circumstances surrounding that buy:

Deane Johnson, KOMA, 1962. Johnson was a program director and on-air talent for Storz stations in Oklahoma City and New Orleans. He also held an F.C.C. First Class engineer's license, which allowed him to operate and maintain transmitters like the high-powered one at KOMA, Oklahoma City. Its signal could be heard halfway around the world at night.

> Todd Storz had made up his mind to buy WIP in Philadelphia, which would have been another major market station. Then he discovered that he would also be buying the services of an entire,

unionized symphony orchestra — hardly a useful adjunct to a Top 40 station! So he walked away from that deal, even though he had the money. Todd heard that KOMA in Oklahoma City was being considered by Gordon McLendon, whose stations were also running the Top 40 format. So Todd bought KOMA kind of "on the rebound." It was a tough market for him.

In a personal essay looking back over his lifetime, Sampson said that "KOMA was one of those trying and frustrating episodes you 'chalk up to experience'.... In retrospect, I learned a great deal from the KOMA experience about how to manage people and difficult circumstances, and that a person should never give up."

Sampson's comments underscore the fact that the advent of Storz-style radio in Oklahoma City in late 1958 was not the slam dunk, overnight success that had been the norm earlier in the decade. What Deane Johnson describes as "a sophisticated form of competitive one-upmanship" had been organized behind the scenes by Katz, a well-known national radio sales representative firm. Radio sales reps act as intermediaries between companies wanting to advertise a product and radio stations wanting to be those companies' medium of choice, by providing research and other services that help clients to buy and sell advertising. Katz had client radio stations across the United States, but they were primarily in the top 25 markets. Because of his long and deep knowledge of how Storz programming and promotion operated, the Katz firm had recruited Johnny Pearson — the top-rated former morning show host at KOWH, and later program director at WHB — as a consultant to Katz-represented radio stations. His first assignment was Oklahoma City's WKY, owned by the Gaylord broadcast group. Oklahoma City was far from being in the top 25 markets, but the Katz station representative firm had a special interest in seeing WKY succeed: George Katz sat on the board of the WKY Radio Phone Company, licensee of WKY.

Said Pearson in a 1994 interview with Richard W. Fatherley, "Just about this time [1958], it was announced that Storz was coming to KOMA. So I said [to WKY management] 'Okay, you're going to have one chance and you're out. Because when they hit town and you don't change this ... and this ... and this.... I'm sorry, you're going to have the same thing happen to you that's happened in other cities." That is, WKY would soon find themselves at the bottom of the ratings, pleading with advertisers to stay with them.

Pearson's work for Katz transformed WKY from a staid NBC radio affiliate into an aggressively programmed imitation of Storz's WHB in Kansas City. Because Pearson had been program director at WHB, he held a critically important set of "keys to the candy store." He knew how to beat Storz at his own game.

Eight • The Air War in Oklahoma City 117

Storz's initial announcement of plans for KOMA caused raised eyebrows in the radio business when *Broadcasting* reported the following: "A leading independent radio operator's station goes network: a leading network affiliate goes independent — that's the radio story in Oklahoma this week." Storz had indeed affiliated with the NBC radio network after WKY took Pearson's advice and dropped the contract. That arrangement was shortlived, however. Soon, KOMA would also drop NBC.

Meanwhile, WKY had been busy researching every successful stunt, promotion, and audience building Storz strategy since the early days of KOWH, including WTIX's introduction of the Top 40 "countdown" format; WHB's whirlwind 1955 treasure hunt in Kansas City; and the twin treasure-hunt giant jackpots split between KOWH and WDGY, which had caused raised eyebrows at the FCC. Like detectives, WKY and KATZ examined every promotional case history relating to Storz success stories. As Deane Johnson summarized it, "No stunt was left un-turned."

WKY program director Danny Williams remembered it this way: "We heard Storz was going to buy KOMA. Storz was one of the real legends of the business. So, what we did was get every promotion that we heard Storz did. We did all of them before they got here. When Storz got here, there wasn't anything left for them to do!"[2]

Storz took control of KOMA on November 20, 1958, but found that the major elements of the Storz Top 40 format were already airing on WKY. Katz consultant Pearson recalled, "I don't think KOMA ever got top position." (KOMA did achieve top ranking once, though only briefly.) Perhaps because of Pearson's long and deep knowledge of how Storz programming and promotion worked, WKY's ability to "beat Storz at his own game" was big news nationally. Pearson's success with Katz soon caught the attention of ABC Radio. "We did the same thing with ABC for various stations we were sent to around the country — with varying results," Pearson said.

While WKY had successfully upstaged Storz locally for the Top 40 audience, KOMA's powerful and unusual signal coverage pattern made it a winner outside the Oklahoma City metro area. In the daytime, the Storz station's three-tower antenna system was driven with 50,000 watts, and even though its 1520 frequency at the high end of the AM spectrum usually spelled weak coverage, KOMA's daytime pattern reached Wichita to the north and Dallas to the south. Its daytime coverage to the east and west included virtually all of central Oklahoma, except the western panhandle. So KOMA had an admirable daytime signal. But after sundown, KOMA's nighttime coverage pattern literally "skipped" across the country into the homes and cars of Top 40 fans all over the region. That was because, after sunset, the directional antenna system was operational, concentrating coverage to the northwest, west, and southwest of Oklahoma City — and to points halfway round the world when the signal was reflected by ice crystals in the ionosphere on a cold night. From as far away as Vietnam, Guam, Alaska, and Australia, mail poured in addressed simply to "KOMA in Oklahoma," a frequently used phrase on-air. Thousands of people in the western states heard Top 40 radio as they had never heard it before. And the nighttime KOMA signal became a goose that laid golden eggs for start-up bands and music groups with names like the Fabulous Flippers and King Midas and the Mufflers. They ran their commercials on KOMA, and occasionally featured a KOMA personality as part of their live shows.

KOMA's original studios were in a downtown Oklahoma City building, but during Storz ownership, the studios were moved to the KOMA transmitter site at nearby Moore, Oklahoma. KOMA air personalities who operated the station solo were required to have an FCC First Class license, to assure that the high-powered transmitter was not interfering with other stations on the same frequency. Shortly after Deane Johnson arrived in Oklahoma City in 1961 to be KOMA's new program director, the station received an unprecedented notice of violation from the FCC. It accused Storz Broadcasting Company of willfully and repeatedly transmitting KOMA's signal at a power greater than its licensed 50,000 watts, and of failing to employ its after-dark directional antenna system between midnight and sunrise — which would have protected other stations broadcasting on the 1520 frequency. In fact, KOMA's high-powered non-directional signal caused late-night interference with the 50,000-watt signal of WKBW, which shared the 1520 frequency more than 1,000 miles away in Buffalo, New York.

The cost to Storz would be a $10,000 fine (about $75,230 in 2012), which in those days was the largest of its kind ever levied against a radio licensee. The FCC had previously permitted broadcasters to use higher power for post-midnight "maintenance and testing authority." That was the excuse Todd Storz used for giving KOMA management the go-ahead to operate the transmitter after midnight without using its directional antenna system. In Johnson's opinion, the violation of operating "far in excess of the authorized 50,000 watts" was not deliberate. Rather, Johnson believed it was the result of aging and long-ignored antenna equipment that had shifted in its electrical characteristics over the years. Storz paid the FCC fine on condition that the words "willful and repeated" be removed from the language of the citation.

WKY's preemptive leap out of the starting gate had given that station time to stake out the classy, "full service" Top 40 position in the market. WKY talent was well chosen and stable, exemplified by their strong program director and morning-show talent, Danny Williams. They also had a fully staffed and

Eight • The Air War in Oklahoma City

credible news department. Several members of WKY's air staff also were featured on Oklahoma City's WKY-TV. Like Storz's WHB in Kansas City — after which they were patterned — WKY was formidable. As a consequence, KOMA was relegated to a position of trying to make a lot of noise and excitement to get attention. In that effort, KOMA had one advantage over WKY that is often overlooked by historians. That advantage was the use of PAMS station identification jingles, and especially the "Yours Truly KOMA" theme that became its trademark. Deane Johnson believes the PAMS jingles (which Storz made sure were offered to his stations as soon as they were released) gave the station an exciting personality. Because Storz very wisely purchased nearly every product PAMS created, the competitors — including WKY — were shut out of that vital programming ingredient.

KOMA had also been provided with the usual Storz station technical tools, many developed by Storz personnel at KOWH in the early days. There was the Hammond Organ "reverb" unit, used by KOMA behind all programming and for emphasis in newscasts. Other devices included the Mon-Key — a short wave code sender modified for use as a news sound effects machine, and the MacKenzie rapid-cueing program repeater, used for jingles and news sounds.[3] KOMA pioneered solid-state audio processing nearly a decade before it became commonplace in the broadcast industry in the form of the Limpander, a combination audio loudness limiter and expander. The unit had been designed by a local electronics engineer to help his special-needs daughter hear better, but he built a unit for KOMA to experiment with. It was not suitable for broadcast use as delivered, but KOMA engineers modified it to achieve unusual clarity and unprecedented loudness on the AM band, something that had not previously existed. It was later adopted by some of the other Storz stations.

KOMA had something else going for it. Now owned by Todd Storz, the station virtually had its choice of up-and-coming talent. Any applicant who got a call from KOMA often became so excited they sometimes forgot to ask if they would be paid for working there. Danny Williams, longtime program director at WKY, recalled that when his station and KOMA were battling it out, they were continually amazed at the quality of talent KOMA attracted. Williams said WKY felt they were in danger of being beaten by KOMA at any time, so they never dared to drop their guard.

Several noteworthy KOMA air personalities established national reputations after accepting high-dollar jobs elsewhere. One of those KOMA alums was Rod Roddy, who was recruited by ABC's owned-and-operated KQV in Pittsburgh — the first network-owned radio station to adopt Top 40 as its program policy. He was "Hot Rod" Roddy, who reported to work in "scrubs" as

KQV's teen appeal air personality from 9 P.M. to twelve midnight. Later, CBS hired Rod Roddy as Bob Barker's announcer on the long-running daytime TV show *The Price Is Right*. KOMA's "Chuck Dann" was actually the voice of Chuck Riley, whose later claim to fame was voicing many of Hollywood's dramatic national movie commercials. He realized enough income from that endeavor to buy the former Hollywood home of silent movie star Charlie Chaplin. And another Charlie — KOMA's deep-voiced "Charlie Tuna" — found fame and fortune at KHJ in Los Angeles. His real name was Art Ferguson, and he came to KOMA from Wichita, KS.

Once past the opening salvos under new Storz ownership, KOMA did not counter-program or react much to WKY — paying only moderate attention to what the Gaylord station put on the air. Whereas WKY's presentation mirrored the more laid-back, classy WHB approach that Johnny Pearson had learned at that Storz station, KOMA had a tendency to inject far more fun and excitement into the sound. On the other hand, WKY had a big news department whereas KOMA had none. To compensate, KOMA pulled news items from the Associated Press printer and punctuated the stories with production sounds and music to make it feel exciting and different. Many other Top 40 stations did the same. News presentation excepted, most of what KOMA did was original; it wasn't copied from any other station or source. Specialized KOMA programming came from the minds of the people who were there, not from a "playbook" of any kind.

Promotions were vital in the battle with WKY. Deane Johnson recalls:

> One of my favorites was actually a WKY promotion that we stole from them right "in broad daylight." WKY Program Director Danny Williams had obtained his private pilot's license and created a stunt for the week when Hooper ratings were taken that involved throwing baskets of cash out of an airplane over certain Oklahoma City intersections. Listeners were to learn at which intersection, and the time of the drop, by listening to WKY. We immediately hit the air on KOMA with our own promotion titled "Money Is Falling From The Sky To KOMA listeners!" The only difference was that "our" money was actually WKY's! By monitoring WKY for the drop announcement, we immediately made our own announcement, making it sound like our own. WKY even tried to make the drop in the middle of our newscasts, thinking we couldn't respond. When we interrupted those, they ended up canceling the promotion. Those were the fun days.

Following the 1958 creation of NASA, plans were initiated to put a man on the moon. That goal was among John F. Kennedy's objectives during his short-lived administration. KOMA eagerly capitalized on the sudden public interest in space travel with a promotion centering on the theme "Win a Trip to the Moon or $500!" Promoting winning a trip to the moon—which until

Eight • The Air War in Oklahoma City

recently had seemed an impossibility—was a big deal. The promotion helped KOMA to seem "bigger than life." But Deane Johnson was counting on the winner taking the cash, and fortunately they did.

Johnson admits that KOMA begged, borrowed or stole promotion ideas from wherever they could get them. Contests aired by other stations, such as "Hi-Lo," "Lucky License Number," and "Name-It-And-Claim-It," all ran in due course. One Hooper rating week, KOMA gave away a copy of every record they played. The record companies were glad to supply the additional discs.

Because the KOMA promotional budget was very limited, it was necessary to invent promotions that didn't cost anything, like the memorable "KOMA Kissing Tone" still talked about today. The wording was: "If you're a loyal KOMA listener, at the sound of the 'coma' Kissing Tone, kiss your sweetheart"—which was followed by an exotic, up-close, dynamically enhanced, and *extremely* long kissing sound. It ran for years.

Three factors reduced KOMA's ability to compete with WKY. One was the remoteness of its studios and offices out at the transmitter site in Moore, Oklahoma. As had been the case with WDGY, the cost-savings realized by giving up in-town studios and offices had left KOMA short on studio facilities. But the federal government suddenly and unexpectedly came to the station's rescue. Cold War worries were responsible for the construction of unusual "bomb shelter" studios at KOMA, and at selected other radio station sites around the country. The new studios were built with very thick concrete ceilings and walls. The federal government specified the design, then paid for construction, new audio equipment, and a huge generator with enough diesel fuel buried underground to keep KOMA running at its full 50,000 watts for a month. The end result was that KOMA gained a much-needed production studio and massive stand by power. Management was encouraged by the government to use the facility at all times so that it would remain ready, and the station was only too happy to comply. When severe weather bore down on Oklahoma City—a metropolis very much in "Tornado Alley"—the new 150,000-watt Caterpillar diesel generator would be started up so that the station's signal would not be affected by any power losses during the storm.

A second factor—which Deane Johnson describes as one of the more illconceived ideas Storz Broadcasting introduced in the early 1960s—was a central music play list, prepared at the home office in Omaha, to be followed by the various Storz stations. The prescribed list put KOMA at a sizable disadvantage against WKY, which was free to play locally popular hits and to tailor their programming to the Oklahoma City market. Deane Johnson was able to convince Jack Sampson that KOMA should ignore the Storz edict and play its own list of music.

The third factor was program automation. A significant action that Johnson took at KOMA — and perhaps the boldest (or maybe the most personally dangerous) — was unplugging the Schafer program automation system, which had been installed early in 1961. In spite of making many radical technical improvements to the equipment, an automation system playing music and voice announcements from big one-hour tape reels, and attempting to integrate them with jingles and commercials and a singing clock, just could not have the spontaneity of live broadcasting. During a visit to KOMA, "Bud" Armstrong — a man who in Johnson's view could have written the bestselling book *Winning Through Intimidation* — asked then-general manager Rex Miller to accompany him to dinner. Miller, being just a bit nervous about having "the man" in town, asked Johnson to join them. The trio had no sooner gotten seated when Bud rather gruffly asked, "Just how automated are you?" Johnson watched the color drain from Miller's face as he slid down in his chair, hoping not to be looked to for an answer, so Johnson responded without hesitating: "Not at all. We're 100 percent live." Armstrong rather firmly followed up with, "And who made that decision?" Johnson could hear Rex gasping for air as Deane responded, "I did." Armstrong replied, "Well, it's about time somebody did. Automation was the worst idea Todd ever had." And that was all that was ever said about dropping automation at KOMA.[4]

A short time after Jack Sampson was promoted from KOMA to manage KXOK in St. Louis in 1964, KOMA finally did beat WKY in a Hooper rating, taking the 9 A.M. to noon and the 3 to 6 P.M. "afternoon drive" day parts. Johnson recalls that "WKY came unglued." Lee Allen Smith, the manager of WKY, called him and wanted to have lunch. Johnson agreed to meet, and Smith started probing him about what Johnson would like to do in the future, because they wanted to get him out of town. Smith plainly said that since Johnson was the jock on the air during the ratings-winning morning day part, it would serve as an inspiration to WKY's other jocks and it would be worth something to them for Johnson to leave. But Johnson made clear to Smith that he had no interest in leaving Storz Broadcasting and the lunch ended. Meanwhile, Dale Wehba — one of the disc jockeys KOMA had bested — was moved to WKY's all-night shift as punishment for getting beaten. When KOMA learned of this, Johnson contacted Wehba and ended up hiring him for afternoon drive, which further "unglued" WKY. Several years later, he became KOMA's program director.

Johnson remains nostalgic for those days at KOMA. He still has the handwritten note from Jack Sampson congratulating him on finally beating WKY, even though, as Johnson acknowledges, "It was a one-time event." Johnson adds:

Eight • The Air War in Oklahoma City

We were young back then. It was nothing for us to pack up and move on to another promising job opportunity. But, for people who worked for Storz ... it was a different world than they'd ever seen before, or since. We were totally dedicated to the Storz Broadcasting Company. That's all we cared about. We didn't care about any other radio station. We didn't care about where we'd go in the future. It was only Storz Broadcasting and winning the next ratings period. But then Jack Sampson was transferred to manage KXOK in St. Louis, and just a few years later, Todd Storz died at age 39. With his passing, the *esprit de corps* of being part of the Storz radio group began to disintegrate.

Following his early 1960s tenure with Storz Broadcasting at KOMA, and at WTIX later in the decade, Deane Johnson took programming assignments at KDWB, Minneapolis–St. Paul against Storz's WDGY (as recounted earlier), and in Cleveland for NBC. But Johnson avows that

> I never enjoyed any of those later responsibilities as much as I enjoyed working with Jack Sampson at KOMA. Jack had the unique ability to be in complete control, yet fade into the background to allow his team to exert creative freedom. Never in my entire broadcast career did I ever work for a better manager. He had our complete respect and admiration.

That is the kind of accolade one might expect to hear from the junior officers when the general is being transferred to a new command. For Deane Johnson and Jack Sampson, the Oklahoma City "air war" had ended in a stalemate. But what awaited Sampson in St. Louis as the new manager of KXOK would be nothing less than the thrill of victory.

• NINE •

The Last Hurrah: KXOK, St. Louis

In hindsight, the 1957 sale of KOWH in Omaha — where the Storz radio revolution had its start — marked "the end of the beginning" for Storz radio. KOWH's sale price of $822,000 (about $6,618,000 in 2012) was then a record for a daytime-only AM station. It reflected the strength of Storz programming and promotion practices, which the new owners could neither emulate nor maintain. It was the only station Storz sold during the heyday of the Top 40 era.

Just three years later, the 1960 purchase of KXOK in St. Louis would turn out to be Storz's last radio station acquisition prior to Todd's untimely death in 1964 — just a few weeks before his fortieth birthday. KXOK would become a hugely successful station, whether measured by program innovation, audience loyalty, or advertising revenues. But it would be the *last* one — "the 'beginning of the end' of the glory days of Storz broadcasting.[1]

Storz had long had his eye on St. Louis, particularly since his great success at WHB in Kansas City, on the other side of Missouri. Several small stations were available, but with their limited power he felt he could not compete with KMOX — or anyone else. KMOX featured "At Your Service" programming — the nation's first full-time all-talk format. The CBS owned-and-operated station had long dominated St. Louis during the day, and could be heard in much of the Midwest over its clear channel at night. In St. Louis radio, KMOX was "the 800 pound gorilla." No other station came near to challenging its supremacy. An additional problem was that in 1960, several other St. Louis stations were already offering the Top 40 format. However, Storz felt they were not doing it well enough, and that their facilities were uncompetitive.

Storz decided to follow-up on a rumor that Elzie Roberts might be in the market to sell his KXOK, a 5,000-watt fulltime station on 630 kHz. In spite of it being a strong facility technically, KXOK's ratings were low, and the station seemed unable to develop programming that attracted strong audience numbers. In addition to KXOK, Roberts owned the *Saint Louis Star Times* newspaper, a distant competitor to the *Saint Louis Post-Dispatch*. As noted by St. Louis radio historian Frank Absher, writing in the February 17, 2009, edition of the *Saint Louis Journalism Review*, Roberts was nervous about the inroads television was making on radio advertising revenues, and was anxious to sell KXOK. Said Absher: "his heart was not in the business. When he was approached by a potential buyer, he was eager to talk."

In late 1960, a deal was struck, for a reported $600,000 (about $4,559,574 in 2012), although other accounts say the price reached seven figures.[2] As part of the contract, Storz agreed to continue to employ Chet Thomas as general manager for a "reasonable [but unspecified] length of time." Thomas also had the responsibility of managing Roberts's KFRU, halfway across the state of Missouri in Columbia, which was obviously a considerable distraction. The "reasonable length of time" eventually became *un*reasonable for Storz; Thomas would not finally retire until 1964. But Storz now had a powerful, low-end-of-the-dial AM signal in the nation's (then) eighth largest market. It was the highest price the company ever paid for a station.

With six operations now in the portfolio, there was talk about picking up a seventh, which in those days was the FCC limit: no more than seven AM stations (and seven FM stations, for a total of fourteen) could be held by any single licensee. WIP in Philadelphia was considered for acquisition by the company, but that buy never happened. Jack Sampson explains why not:

> WIP was owned by Bennet Gimbel, of the Gimbel's Department Stores chain in New York. Todd was very hot for the station. If he got it, Bud Armstrong would go to Philadelphia to manage it, with either Steve Labunski or me managing WHB, and the other guy going to WDGY to take Labunski's place. But at their final meeting ironing out the details, Gimbel informs Todd that he had a complete symphony orchestra on the staff of the station! They were all union members, and he had just signed a three-year extension on their contract! So Todd would have to take over that obligation. He balked, and the whole deal fell through. But Todd had that money ready to go, and that's when he bought KOMA in Oklahoma City — kind of "on the rebound" because he couldn't get WIP. Storz also looked at buying WINS in New York, but he didn't want to pay the asking price. And because of his father Robert's antipathy to FM stations, Todd never acquired FMs in the cities where they already owned AM stations.

Neither Sampson nor anyone else in Storz management knew it at the time, but the acquisition of KXOK would be Storz Broadcasting Company's "last hurrah."

KXOK was an excellent facility technically. With a daytime signal covering east-central Missouri and central and southern Illinois, KXOK blanketed more than 52,000 square miles, touching the western edges of Indiana and Kentucky. Like WQAM in Miami, 630 KXOK was a low-frequency AM station with a full-time power of 5,000 watts. Its three-tower antenna system was planted in the moist bottom-land across from St. Louis on the Illinois side of the Mississippi River, where water in the soil enhanced electrical grounding and thus increased the station's coverage. At night, KXOK's directional signal kicked in and headed straight south down the Mississippi like a blowtorch, with one engineer exclaiming, "We burn toast in Memphis!"—a figure of speech often used to brag about a station's high signal-strength. There was every reason to believe that under Storz management, KXOK would soon be on the way to massive success.

But at the same time Storz was acquiring a fine facility in St. Louis, his family life was beginning to fall apart. His wife of fourteen years filed for divorce on January 5, 1961. The *Omaha World Herald* reported the $425,000 settlement (about $3,197,293 in 2012) in its January 23, 1961, edition: "In addition to cash, Elizabeth Ann Storz received the household goods, her jewelry and clothing, a 1957 Cadillac, a savings account in a Minnesota bank, five hundred shares of stock in an oil corporation; and custody of their daughter, 11 year-old Lynn Ann."

A month later, Storz announced that the company had purchased a building in Miami Beach to be "extensively remodeled and completely redecorated" as the company's national headquarters. A modern, state-of-the-art recording studio was also to be built so that professional recordings by top artists could be produced there.

In retrospect, there were several good reasons for Storz to move his home office from Omaha to Miami. Doing so would distance him from his failed marriage, remove him somewhat from his father's close scrutiny, provide him with a fresh location more in keeping with a modern, upbeat corporate identity, offer hoped-for relief from a persistent sinus condition and migraine headaches, and bring him closer to the woman who would become his second wife, Helen Lorraine Smith, an attractive blonde from Wisconsin, then WQAM's receptionist.

KXOK's Chet Thomas made a poignant observation in his memoir: "There was something mysterious about Todd Storz. He was a loner and kept people at more than arm's length. I had no quarrel with this, except it did get irksome when an important decision had to be made, and there was no way to reach Todd. Everything was filtered through Bud Armstrong." That is the way Storz wanted it in those days; let Armstrong run the company so that Storz could enjoy his life.

But the "filtering through Bud Armstrong" Thomas complained of was not the case in recruiting a new program director for "The New KXOK." For the first few months under Storz ownership, KXOK had languished, with its ratings under five percent. Storz made it his personal business to re-recruit Bud Connell to "fix" KXOK's programming. Thus, in June of 1961, Connell found a telegram from Storz under his front door. The message was short, and to Connell, sweet. It offered Storz's largest station, and nearly double his income, if he would rejoin the company.

As outlined in previous chapters, Bud Connell had given Todd a major run for his money, by being the first broadcaster to decimate Storz's audience in not one, but two major markets, through the use of tactics he had learned from Storz. He had been hired originally in early 1957 as a morning personality at Storz's flagship, KOWH, just prior to its sale. Disillusioned that his contract was sold along with the station, Connell soon left Omaha to program WNOE and thus compete head-to-head with Storz's WTIX in New Orleans. Nine months later, WNOE was the top-rated station along the entire Gulf Coast. Storz personally interviewed Connell about managing WDGY in Minneapolis but ultimately decided

Bud Connell, KXOK, 1968. Connell did not invent the Top 40 format; he just made it more fun, and far more profitable. A master programmer, Connell beat Storz's WTIX in New Orleans and WQAM in Miami. To keep from being bested a third time, in 1961 Storz hired Connell to program KXOK in St. Louis. It became the top-rated station in the city, and the number 1 independent station in America. Connell asked Todd's father, Robert H. Storz, to consider acquiring an FM station in St. Louis in order to stay competitive, and was rebuked.

that Connell — then only 23 — was too young for the job. Connell subsequently teamed with Robert Rounsaville to build and manage the new WFUN, Miami, deflating WQAM's South Florida ratings in only three months, as described earlier. Marc Fisher reports in *Something in the Air* (p. 22) that Storz sent Connell a four-word telegram: "Okay, I give up." It was the first step in Storz's effort to re-recruit Connell to program KXOK. When Connell asked, "How much is my operating budget?" Storz's answer was short and sweet: "You don't have a budget. Spend whatever you need to spend, but spend it wisely." That last phrase probably felt to Connell like a father's admonition, but he was only 25 years old at the time.

Connell's initial task would be to monitor, evaluate, and re-program KXOK, and to replace personnel as necessary. Three existing air personalities were given notice so that Connell could bring in several of his best performers from New Orleans, Miami, and other areas of the country. Among them was Danny Dark, who Connell lured away from his former employer, WFUN. Dark would ultimately become one of America's most familiar voices — as the NBC television network's chief announcer.

The objective at all Storz stations was to build a strong relationship among the listener, the station, and advertisers. About one-third of all local commercials were delivered "live"— they were read or ad-libbed by the person on the air. Local advertisers liked the individualized attention they received from a Storz "air personality." In the background, Bud Connell was creating KXOK's "station personality" which delivered a growing audience to those advertisers. In turn, burgeoning audience growth allowed KXOK to raise advertising rates, with the result that KXOK was on the way to becoming not only the most listened to independent (that is, non-network) radio station in the nation, but also one of the most profitable. Connell was promoted to station operations manager, and he assigned morning personality Ray Otis as his liaison program director with the air personalities and the news staff.

When Connell had arrived at KXOK in July of 1961, the station attracted just 4 percent of the audience. Top 40 competitor WIL had 20 percent. For the next several months, Connell made changes that produced the largest impact for the least effort, while he and his new staff wrote and recorded a completely new format: promotions of all types, energetic introductions and closes for news and all other features, exotic/parody/lampoon commercials, new custom singing jingles and musical "punctuators,"[3] a wide variety of audience effects, and an array of running humor bits. KXOK's entire new format differed from what was heard on the other Storz stations because it was a hybrid melding of the elements Connell had created to steal Storz ratings in his previous jobs. Connell is quoted in Marc Fisher's *Something in the Air* summing up the appeal of the Top 40 format that he had mastered so well: "The context was always predictable and the content was entirely unpredictable" (p. 342). That looks like a simple aphorism, but it is a distillation of years of experience and insight.

The new sound of KXOK was introduced on October 1, 1961. The format redesign included the slogan "The Spirit of St. Louis"—a smart nod to the city's pride in the nine St. Louis businessmen who had helped finance Charles Lindbergh's solo transatlantic flight in an airplane with that name. In the afternoons, Ray Otis hosted a hybrid version of *The New WTIX* countdown show on KXOK. Connell remembers referring to it as "the ladder show." The idea was to build anticipation. It began at 3:00 pm with the top 10 played in reverse order, followed at 4 with hits 30 through 21, performed the same way. Beginning at 5, hits 20 through 11 were played, and at 6 there was a repeat of the top 10. The KXOK countdown program was called *The Top 6 Plus 30* to underscore KXOK's AM dial position at 630 kHz. The four-hour program included two alternating hourly "pick hits," records that had not yet made their way onto the charts.

Note that the number of records typically being played in 1961 had been reduced from 40 to about 30. *Variety* had detected the trend of further pruning the Top 40 playlist as early as February 1, 1961. An article titled "Music Biz in Formula Bind" noted,

> The straightjacket of formula radio is now pulling tighter than ever on the music biz. Heavy stress of the vast majority of radio stations on programming the top hits has created an imposing wall against new material as well as causing a quicker than normal exhaustion of the bestseller.
>
> Whereas a couple of years ago, the radio outlets were covering the Top 40 or top 50 songs, currently the number of featured hit songs are shrinking down nearer the top 30 mark. Under a typical formula now used by stations, the top 30 numbers are being supplemented by a half-dozen new songs each week plus an equal number of new albums for the full pop programming fare [P. 57].

In sum, by early 1961 the limited playlist had become considerably more limited, with most Top 40 stations playing only 30 hit records or fewer. In a later interview, Bill Stewart identified the reason for repeating only the top hits when he was programming Storz's WQAM in Miami: "When we first went into Miami ... we went in with, I think, 25 records. We never would admit it. I don't think it ever got over really 30 records. We would *print* a Top 40. But we only played 30. And it worked. Because, again, it eliminated chance—that's the only thing it did."

Records in that era had gotten shorter, too. The record companies realized that a short record stood a better chance than a long one of making it onto a station's Top 40 list simply because it left more time to play other program elements. For example, the early '60s hit by Little Jo Ann titled "My Daddy Is President"—supposedly Caroline Kennedy singing about her father John F. Kennedy—ran less than two minutes.

In October of 1961, the Radio Advertising Bureau (RAB) announced that the average daily radio audience had increased 15.6 percent in the four years since 1957—from 68,354,000 to 79,003,000—while the U.S. population had grown only 6.2 percent in the same period. Although the growth in the average daily radio audience could not be attributed directly to Top 40, the 1957–1961 time span was the period in which the format spread virtually throughout the radio industry. That growth also corresponded with the advent and evolution of portable transistor radios which allowed radio listening almost anywhere.

When the November–December 1961 Nielsen ratings appeared, KXOK had shot from a 4 percent to a 20 percent share of the St. Louis audience, reversing its position against WIL after just 90 days of the new programming. By July of 1962, KXOK was rated number 1 in St. Louis by *Pulse*, a "weighted" survey. (The station's ratings would continue to grow, allowing KXOK to declare in the mid–1960s that according to *Pulse*, it was the *nation*'s top-rated independent station.)

Although he now had impressive new audience numbers to sell to his advertisers, General Manager Chet Thomas had problems translating high audiences into revenue—because he was conditioned to selling long-form network-style programs, rather than the spot advertising that had become the norm on music-and-news stations. At a meeting of all the Storz Broadcasting Company station managers in St. Louis in January 1962—a year and a half after acquiring KXOK—Storz called Jack Sampson aside. (Sampson was then vice president and general manager of KOMA in Oklahoma City, and before that had been sales manager at WHB.) Storz told Sampson that he wanted to transfer him to St. Louis to be vice-president and general manager of KXOK, "as soon as we can solve the Chet Thomas situation." It was becoming obvious that Thomas—even with his deep roots in the St. Louis community, and as fine a gentleman as he was—either did not agree with, or could not understand the Storz operation, the programming, and how to sell it. Thomas was accustomed to selling specific shows and features. Instead, Storz urged his managers to "sell audience," and steady increases in audience was what Bud Connell was delivering. That frequently put Thomas at odds with Connell. Thomas simply did not like the format, and apparently could not figure out how to build a sales staff that knew how to sell advertising on a Storz station. Nor could the KXOK sales staff hide the fact that they were loyal to Thomas. But the agreement to continue Thomas as manager for "a reasonable length of time" probably had not been met yet. Jack Sampson would remain at KOMA for another 18 months.

By early 1964, it was clear that KXOK's financial situation was critical, its programming and audience successes notwithstanding. The station was losing thousands of dollars each month, and was further crippled by three labor union contracts and by old equipment that impaired the Storz sound. In April of that year, Bud Armstrong promoted Connell to the position of KXOK station operations manager, and finally transferred Jack Sampson to St. Louis to "put the bottom line" on the station's profit-and-loss statement as KXOK's new general manager. When Sampson asked for more detail about what Storz wanted him to do, Storz's only reply was: "Make a profit and don't bother me."

Sampson admits that it had taken him a while to learn how to be a station manager at KOMA: "Oklahoma City was my crucible. I had a lot of problems there. But when they promoted me to manage KXOK in St. Louis, things just turned around." Sampson is quick to credit Bud Connell for the upturn in KXOK's fortunes, adding Todd was involved in those days. When I got there, KXOK might have been a tiny bit ahead of WIL, which was

Jack Sampson, KXOK, 1966. Sampson was a salesman at WHB before Storz acquired the station, and became sales manager there in 1956. In 1958, Sampson became manager of KOMA, Oklahoma City, then in 1964 was named general manager of KXOK, St. Louis. He was glad to support Bud Connell in making KXOK the top rated independent (non-network) station in the nation. Like Connell, Sampson also urged Robert H. Storz to acquire an FM station, and was ignored.

also doing Top 40. They had some top talent and some very good kids on the air. We just kept eating away at 'em and finally beat 'em. WIL changed programming and went down the sewer. But they were very competitive for a while."

Jack Sampson and Bud Connell became an immediate mutually supportive team. Sampson later commented, "All I did was believe in the format and support Connell in every way possible." But urged to say more about his own work, Jack admitted that he "completely rebuilt the sales staff, and created a marketing plan to enhance our growth."

On the strength of Sampson's sales and management know-how and Connell's programming leadership, KXOK billings grew in succeeding quarters. In the early months of his tenure at KXOK, Sampson felt that he should spend as much cash as possible on equipment upgrades, personnel quality, and promotions. That strategy began to pay off immediately. In 1964, sales doubled over the preceding year, and income continued to rise.

With achievement of top audience ratings in the record books, Bud Connell plowed ahead with Todd Storz's blessing — and an unlimited promotion budget. The object: hold onto the top St. Louis station position, build a listener base that would dwarf all direct competition, and focus on decimating what remained of the younger audience of the decades-dominant KMOX.

To meet those ends, Connell focused on voices — of two kinds. First, he brought truly "huge" adult voices to the air, both as program hosts and as newscasters. Richard Fatherley was one of them. He became KXOK's production director in April of 1964. For a time, Fatherley was also the "KXOK Millionaire," handing out money to total strangers while riding around town in a stretch limo. Another was Ray Otis, who was named program director. Robert R. Lynn was a first-rate reporter and newscaster. Steven B. Stevens and David D. Rogers both had bass voices that could rattle windows. Even if KXOK was accused of playing music for "kids," the majority of the people on KXOK's air were undeniably mature.

Except when they weren't. The creation of station-owned characters became Connell's next focus. First to be born was a character whose legendary status in St. Louis radio persists a half-century later: "Johnny Rabbit." To fabricate the name, Connell simply melded elements of the two trendy, talked-about items. With "how can it miss" reasoning, he took TV talk show host Johnny Carson's first name and welded it onto the word for the *Playboy* logo. Johnny Rabbit was an instant success, in part because of the vocal talent and creativity offered by Don Pietromonaco — who was Rabbit's voice — and in part because in his evening time slot, he dominated teenage tune-in. Storz had the foresight to file for a registered service mark with the U.S. Trademark and

Patent Office on the name Johnny Rabbit. That turned out to be important, because the first guy who had been hired to be Johnny Rabbit — Ron Elz — did not work out and was let go. He was replaced by Pietromonaco, who became the definitive Johnny Rabbit on KXOK. But then Ron Elz was hired to be Johnny Rabbit by Top 40 competitor WIL. KXOK got a restraining order to keep him off the air. The case went to federal court and was decided in KXOK's favor. KXOK manager Jack Sampson remembers the infringing disc jockey walking up to the judge and saying, "Judge, can't I be Johnny Rabbit anymore? And the judge glowered at him and thundered "You can't be Johnny Rabbit, Peter Rabbit, Tommy Rabbit, any other little furry animal, or a colorable imitation thereof!" But eventually, Storz abandoned the registered service mark, and Elz acquired it after all.

Another key character voice was KXOK's Kay, who was Rabbit's constant companion when he was on the air. She functioned as the conduit between Johnny and his listeners, and her name helped reinforce call-letter recall. Johnny would chant, "Call the station and blab it to the Rabbit" or "What do you know, what do you say, call Kay." And they did. Approximately 5,000 calls per night were processed by five young women who each fit the Kay profile on the phone. They copied messages and requests from listeners and ferried them to The Rabbit, took votes on KXOK's

Dick Fatherley as the KXOK Millionaire. At KXOK, Dick Fatherley was not only production director (work in which he used his deep, mellifluous voice on commercials and promotional announcements), but he also played the role of the KXOK Millionaire. In that guise, he was driven around St. Louis in a limousine, giving away KXOK cash, and autographing photographs like this one.

"Make It or Break It" new-record feature, and became part of the parade of memorable voices that contributed to KXOK's success.

But not all of the "specialty" voices were aimed at teenagers. Air personality Peter Martin, played by Jim Irwin, was designated KXOK's Poet Laureate, and once per morning on weekdays between nine and noon, he mellifluously performed brief poetry selections for his largely female audience. Ray Otis, KXOK's morning man and program director, hosted several personalities who became KXOK mainstays. Among them were Lou Cooley, a champion boat racer, who was KXOK's morning traffic reporter, and St. Louis's mayor, Alphonse Cervantes, who spoke by "radio-telephone" each morning on Ray Otis's show as "hizzoner" was being chauffeured by limousine to City Hall.

One might wonder how a Storz Top 40 station could get away with such deviations from the prescribed formatting. First, Connell recruited people who could be entertaining without skating on the edge of bad taste. But he was careful never to restrict a workable idea — or even an outlandish one — from the format. His approach was try it, and if it works, keep it. Connell's overriding goal was the melding of individual personalities and fresh (sometimes wild) ideas into that one big "personality" called KXOK. Constructing radio station programming that appealed to all kind of listeners at all times of the day — and night — was one of the ways that Connell accomplished the seemingly unthinkable.

Sampson remembers that

> after Todd died [on April 13, 1964], which was shortly after I got to KXOK, the Top 40 "formula" almost never changed after that — clear up to the end when they were getting eaten up by FM's. A lot of corporations get that way: You get a "formula" that works, and you like it and understand it and live it, but you've lived with it so long that other people are doing things better. If you don't change with the times ... that was a lot of Storz's problem [in the later years of the station group]. After Storz died, they did no more upgrading or buying — especially not FM. But the "formula" worked well into the late 1960s, when the FM's began to get big. And under Todd, GM's had a lot of autonomy. That also changed after Todd's death.

In the summer of 1966, KXOK landed the highly coveted sponsorship of The Beatles' appearance at Busch Stadium, scheduled for August. In July, the *Pulse* ratings service declared that KXOK was the number one independent radio station in the entire United States. WIL needed to stop hemorrhaging audience, so WIL manager John Box hired a high-powered media consultant to "beat KXOK." The consultant believed the "English Invasion" (the craze for any rock 'n' roll band that had an English accent) had run its course, and

Nine • The Last Hurrah

therefore WIL planned a big promotion in St. Louis's famed Forest Park that urged listeners to burn their Beatles records. Jack Sampson recalled how KXOK did just the opposite: "Connell had us really lay on the 'Fab Four.' He hired an English model — Delcia Devon — and George Harrison's sister Louise, to keep us in the British mood" with on-air comments uttered in their quirky British accents. The KXOK pro–Beatles promotion worked so well that after six months, the consultant was gone, and WIL changed its format to country music. Soon, KWK also changed formats and surrendered the St. Louis Top 40 market to KXOK.

In 1967, Richard W. Fatherley accepted a promotion from production director at KXOK, to become program director at WHB in Kansas City — the city where he would remain for the rest of his life. It was also the year when KXOK program director Bud Connell urged Todd's father, Robert H. Storz, to consider acquiring FM facilities, because stereo FM stations were beginning to siphon audience away from monaural AM outlets such as KXOK. But Robert had never gotten over the fact that under his and Todd's ownership, KOWH-FM had been *donated* to the University of Nebraska, Omaha — not sold for a profit. It was the quintessential example of his mantra that "FM" stood for "Failed Money."

A year later in 1968, Connell was fired by order of the elder Storz, in spite of the fact that during that year, KXOK could consistently claim more audience than CBS's KMOX. Moreover, KXOK was the top-billing station in the Storz group, with annual profits in the millions. But a clock was ticking. The days of Top 40 programming on AM stations dominating radio listening were numbered, and the Storz stations' decline would be hastened by Todd's father.

Jack Sampson left the Storz Broadcasting Company in 1975. He had been with them for 24 years. One of the first things he did "in retirement" was to buy an FM radio station.

• TEN •

Elements of the Storz Station "Sound"

During the peak years of their success (from about 1954 to a few years after Todd Storz's death in 1964), the Storz stations had a "sound" that was different from — and better than — any of the many imitators of the Top 40 format.

That unequivocal statement can be supported by listening to "telescoped airchecks," recordings of the station's signal as transmitted, but often with commercials removed or shortened, and usually with the majority of music recordings edited to leave only their beginnings and endings. Such recordings give a "time-compressed" sense of a station's overall sound, and serve as a record of the verbal content that was transmitted.

As of the summer of 2012, telescoped airchecks of selected Storz radio programming can be heard online at the sites listed below.

    KOWH: *www.deanejohnson.net/audio/KOWH_Sandy_Jackson.shtml*. This telescoped aircheck from KOWH in 1957 includes weather, news, a KOWH jingle, Sandy Jackson's theme song, and a few words of Sandy's introduction. This is the only known recording of Sandy Jackson on KOWH.

    WTIX: *www.deanejohnson.net/audio*. When the website opens, use the scroll bar to go down to the WTIX heading, then click on "Deane Johnson Aircheck — 5/10/66." For a 1959–1960 aircheck of WNOE (WTIX's chief competitor), go to *http://www.youtube.com/watch?v=d31412itntw*. This recording was made from a car radio — the reason for the occasional static. The slightly crazed disc jockey at the beginning of the clip is none other than Bud Connell, followed by Shad O'Shay.

    WHB: You can enjoy a 50-second WHB audio and video montage at *www.youtube.com/watch?v=bVRG4uZGrwA*. A five-minute aircheck from May of 1960 is at *http://www.youtube.com/watch?v=fhkDHNt93Fc&feature=related*.

    WDGY: *www.twincitiesradioairchecks.com/wdgy1130tapes.html*. There are many recordings to choose from here, but a November 1956 aircheck features several disc jockeys who read the commercials live (the world's first teenage Top 40 disc jockey Bill Armstrong is among them). Scroll ⅔ of the way down the screen, find the headline "Lunch with Donald K. Martin..." and then click on the "sound" link labeled "WDGY 1130 November 1956."

WQAM: *www.560.com*. In the blue bar on the left side of the opening screen, click on "WQAM." On the orange page with the tiger logos that opens next, choose "Air Checks" in the left-hand column. Recommended shows are April 1962 — Charlie Murdock; Feb 1, 1964 — Rick Shaw; and June 20, 1964 — Jim Dunlap. You can hear a highly dramatic newscast on WQAM's chief competitor WFUN at *www.deanejohnson.net/audio/WFUN_Fundamental_News.shtml*.

KOMA: *www.deanejohnson.net/audio*. Scroll down to the KOMA heading, then click on "Deane Johnson Air Check 3/30/1965." This is an excellent example of music chosen for daytime listeners (i.e., "housewives") rather than teenagers. The great PAMS jingles of the day are heard, and Johnson's announcing is upbeat.

Also under the KOMA heading, click on "KOMA Action Central News with Deane Johnson 2/8/1964." The text included alongside this audio clip gives good detail of how KOMA's programming was produced using MacKenzie Repeaters and an automation system.

KXOK: Go to *http://630kxok.stlmedia.net/*. This site includes video of the KXOK studios (known as "Radio Park"), and offers airchecks by clicking on the blue "Listen Here" text under the small heading "KXOK DJ Shows." Included are recordings of the wacky Johnny Rabbit character.

On YouTube, you can listen to a 1967 newscast with Robert R. Lynn at *http://www.youtube.com/watch?v=DT1HB88BiQU*.

A ticket give-away contest for the KXOK-sponsored appearance of the Beatles is at *http://www.youtube.com/watch?v=FlXwQ0N3Klc*.

One of the many PAMS KXOK jingle packages that were directed by Bud Connell can be heard at *http://www.youtube.com/watch?v=D200KOZffoA&feature=related*.

www.reelradio.com. This site does not have any airchecks of KOWH, but it does offer 11 WTIX airchecks from 1962 to 1972, 8 airchecks of WHB from 1960 to 1970, 8 WDGY airchecks from 1961 to 1968, 17 WQAM airchecks from 1959 to 1970, 4 KOMA airchecks from 1965 to 1991, and 7 KXOK airchecks from 1962 to 1982. Access to most of the reelradio.com archive is by subscription.

There were many contributors to the unique sound of a Storz station. In actual practice, the three most important elements that coalesced to produce the Storz sound were (1) the live and local air personalities (the disc jockeys), (2) live and recorded program elements, such as weather, news, jingles, commercials, etc., and (3) the technical production facilities and quality of the station's sound.

1. Live and Local Personalities

Storz disc jockeys were heavy on personality. They were intended to be distinctively entertaining. But most of all, they were live, and they were local.

They were knowledgeable and enthusiastic about the city and suburbs they were speaking to. Their energy contributed to a feeling of engagement and excitement that is missing from much of today's satellite-delivered hit music programming, which lacks any sense of place.

Former KOMA Program Director Deane Johnson says that the live on-air personality was a mainstay of Storz programming beginning with the early days of KOWH — when Sandy Jackson quickly developed a Hooper rating exceeding 50 percent of the available audience, allowing KOWH to claim more audience than all other Omaha stations combined. Says Johnson, "That early experience with a 'heavy' personality like Sandy Jackson shaped their thinking on the value of personalities forever." (Johnson deserves praise for finding and posting the brief but only known recording of Sandy Jackson on KOWH. It is the first one on the above list of telescoped airchecks.)

Bud Connell also feels that strong air personalities like Sandy Jackson were an essential element of the Storz stations of the mid–1950s:

> In the early days — about 1955–1956 — WHB was, in my opinion, the purest example of the true *developed* Storz sound, whereas WDGY (with Jack Thayer, Herb Oscar Anderson, et al) was the early example of how far individual markets could go with personalities. WHB sounded wonderful in those days, but the aforementioned two men took the Storz Sound to a new level. My exposure to Jack Thayer and Herb Oscar Anderson on WDGY — more than almost anything else — informed my later programming decisions about personalities. Even earlier influences were Jack Benny, Bob & Ray, Bob Pool of *Pool's Paradise* on WWL, New Orleans, and Stan Freberg. Those five performers in the late 1940s and early 1950s gave me the impetus to try my brand of station-wide irreverent humor, which worked like magic wherever I installed it — before, during, and after my Storz affiliation. My best examples were WFUN Miami in 1961, and KXOK St. Louis 1961–63. I highlight the '61–'63 time frame of KXOK because that was the image-establishment period. The first 30 months of KXOK were a wall-to-wall sound carpet of pushed-to-the-edge personalities, outlandish news formatting, and continuous promotions. That brief period so entrenched the station in the minds of St. Louis listeners that we could load up with commercials for almost a decade after, and still remain dominant.

2. Live and Recorded Production Elements

Music: Although the Storz stations had collectively embraced Top 40 company-wide, when Bud Connell opened "the new" KXOK, St. Louis in mid–1961, he employed a "top 50 — plus" approach. Connell explains, "There were already two stations in the St. Louis market identifying themselves as Top 40s — KWK and Balaban's WIL. What does good marketing dictate? Open with something bigger and better — and we did." KXOK never identified itself as Top 40 on the air, and by the mid–1960s, when the competition had given up and changed formats, Connell opted instead to emphasize KXOK's 630 kHz dial position by playing the "top 6 plus 30." As explained in a previous chapter, playing the top 30 hit records instead of 40 reduced the chance of airing a song that would not become highly popular. As was the case with other Storz stations, KXOK also played Pick Hits, Extras, and Golden Oldies.

Choosing the right music for airplay was taken very seriously at Storz stations. Bud Connell explains the process:

> Each record considered for airing was carefully selected from recent new releases. The Top 40 records were determined from national chart positions reported by music trade magazines, averaged with local sales and jukebox play. A few records that were getting airplay on other local stations might also be included, all according to strict locally developed formulae. I used rank weighting in my St. Louis formula: the higher the recording ranked on a given sales or play chart, the lower the numeral it received (number 1 being the lowest). The higher rated the recording, the more exposure it received on the air; conversely, the lower the record rated in popularity the less airplay it received. The number 1 record was played far more often than say, number 27.
>
> The primary element governing individual record choice and exposure (how often it was aired) was popularity with the audience. Weekly chart position reflected that popularity. In practice, the number 1 song was played once every two hours; records number 2 through number 10 rotated every three hours. The record which was designated as the Pick Hit was played once every two hours, or in the case of KXOK, the two DJ Picks of the Week rotated once per hour. Golden Records (or Golden Oldies) were also divided into categories determined by year of popularity and all other codes and restrictions governing current records. Goldens were played once or twice per hour, depending on the station. The balance of the hour was comprised of strict rotations of the remaining categories: numbers 11–20, number 21–up, and Extras, which were usually limited to seven to ten selections. Extras were "newer" records that had not yet "charted" but that were likely to climb into KXOK's top 6 plus 30. Certain records, though, were available only in specific day parts, or specific hours within those day parts because of style "intensity" restrictions. Within a given musical style category, the intensity designations were one ("light"), two ("medium" or "moderate"), and three ("extreme").

The actual airplay formula for the spaced repetition of hit records varied among the Storz stations; however, all stations started with an hourly form (the "clock") that specified the play order of the various rotation categories. KXOK, for example, relied on a form that had to be filled out in duplicate by the air personality prior to his showtime. The show's engineer received the duplicate copy, from which he pulled each hour's music (on tape cartridge) to be played in the listed order. Every day part segment (5 am–9 am — "morning drive"), 9 am–noon, noon–3 pm, 3 pm–7 pm ("afternoon drive"), 7 pm–

midnight, and midnight–5 am ("overnight") specified a different music mix, and the mix also varied from hour-to-hour in each day part — a not-so-easy feat in the years before computer-generated playlists. According to Connell, there were thirteen different day part forms covering the seven-day week, and fourteen combined hourly variations within day parts, each designed to govern pace and tempo in consideration of the average listener's activity and behavior.

The records were labeled with title, artist, and length. They also carried a color code, and a tempo designation. The color codes differentiated each record by style: general pop, teen-appeal, rhythm 'n' blues, country, and album-oriented rock; and by tempo: up-tempo, moderate-up, moderate, moderate-down, and down.

Concerning the tempo of the records played, Connell had a trick up his sleeve, which soon spread to other Storz stations and then to other leading outlets around the country. He directed his engineers to wrap a small length of splicing tape around the spindles of the motors which drove the turntables. The tape-enlarged motor spindles caused the turntables to rotate slightly faster than 45 rpm, thereby giving the record a sprightlier sound. Listeners who dial-hopped among stations, said that the music "sounded better" on KXOK, and seemed to "drag" on the other stations.

Deejays at Storz stations were obliged to follow their playlists, which dictated a smooth undulation of tempo and style throughout the day and evening. Each of the previously mentioned style and tempo categories achieved a degree of prominence at some point in the 24-hour broadcast day. For example, certain selections that had something of a country flavor were restricted to 5 to 8 am, while Album Rock-styled records could be included in the rotation only between 6 and 11 pm and during certain weekend day parts. The core music list of "single intensity" selections of all variety could be played at any time; "double intensity" could be aired at most times; but "triple intensity" selections were greatly restricted — to prevent older listeners from tuning away.

Everyone — including a few of KXOK's competitors — thought their deejays just picked up a pile of records and played them. Even the more alert listeners never could quite figure out why KXOK sounded so smooth compared to the competition. The many ways in which airplay music was selected, analyzed and scheduled at all Storz stations is a key insight into the company's unprecedented success in attracting, holding, and growing massive audiences. It cannot be emphasized enough that it was this level of exactitude and "granularity" in selecting and programming the music that made Storz's brand of Top 40 radio so instantly recognizable and so commercially successful. Most

Ten • Elements of the Storz Station "Sound"

other station operators who played the Top 40 records as listed in music industry trade magazines simply did not make the same effort to assure that the song being aired was appropriate for the market and for the given day part, that it matched up with the available audience, that it worked well with the record that preceded it and the one to be played next, and that it was aired with a frequency consistent with its popularity.

It is incorrect to think that Top 40 radio stations in the 1950s and 1960s primarily played rock 'n' roll records. On evenings and weekends, when teens were most available to listen, rock 'n' roll artists did get a lot of airplay. But during the rest of the week, so did country singers, rhythm 'n' blues bands, ballad crooners, full orchestras, jazz bands, even "chipmunks."

The huge audiences that could be accumulated by a single Top 40 station playing a wide variety of music is unimaginable by today's radio music listener, whose taste has been segmented and forced into ever-narrower "niche" formats. (One genre example: "country" is now split into tightly targeted groupings such as hot country hits, country oldies, country/rock, alternative country, etc.) The consequence of modern-day radio stations attempting to "super-serve" ever-narrower audience tastes is that the elements of surprise and diversity that were present in Top 40's heyday have nearly disappeared. Today's anemic music radio audience figures can only partially be blamed on the rise of online listening and mp3 downloads. Insipid "voice-tracking"—in which an announcer intones the title of the song and the name of the artist (the two pieces of information that are *least* likely to be novel to listeners)—is also at fault. Music on the radio has become a "utility" at the very time when highly rated network television shows are featuring passionate but unknown artists performing their songs to adoring audiences.

Sound Effects: Sound effects—both live and recorded—were elements of the KXOK sound, as well as the other Storz stations. A variety of studio sound effects and props were available for air personalities to use at will. A three-foot tall door in a frame mounted on a stand was custom built and within reach of any deejay who wanted to create the sound of someone entering or leaving the studio. Some of the deejays were multi-voiced talents, and could carry on conversations as if two or more people were in the room, and the door allowed them to enter and leave with a flourish. The live prop was more effective than a recorded substitute, because each live use of the door sounded different. A variety of bells, buzzers, gongs, and other sound effect props made their way in and out of the studios over time.

Borrowing from successful television shows, Bud Connell acquired laugh tracks, various levels of applause, and other short audience sound effects that air personalities could play with the touch of a button—but with the caveat to "use only to enhance a bit that is genuinely funny or worthy of applause." Extensive recorded sound effects libraries were available at all Storz stations for use in commercials. For example, a ticking clock featured throughout a commercial would make the case insistently that "time is running out" to take advantage of the bargains being listed.

Microphone Filter Effects: In the early 1960s, Dave MacFarland worked at a small (non–Storz) Top 40 station where the metal diaphragm from an old earphone had been attached next to the on-air microphone. When it was time to give the weather forecast, he pushed a button which turned off the normal microphone and connected the very tinny-sounding earphone, now operating as a "microphone." The effect was like connecting to a typical low-fidelity telephone line from the weather service, and the highly-filtered voice sounded like somebody else was speaking. Storz stations used the actual low-fidelity mouthpiece from an ordinary telephone handset, but the result was the same: the listener was led to believe that "the weather bureau" was giving the facts.

Reverb Effects: Reverberation is the hollow, spacious sound that is typical of a large room having multiple reflective surfaces. ("Reverberation" should not be confused with "echo" which is a discrete repetition of a sound, that gradually fades out over time.) Some Top 40 stations applied a small amount of reverberation on their studio microphones all of the time, to make the space where the announcer or newscaster was speaking sound as large as possible.

Storz stations also used reverberation selectively. In a newscast, every time a new dateline was announced (for example, "Los Angeles!"), a button could be pressed, and the dateline would have heavy reverberation to make it stand out. The effect was generated by a reverb unit that engineer Dale Moudy originally scavenged from a Hammond electric organ.

Echo effects: Echo is not the same as reverberation. Echo is comprised of discrete repetitions of a single sound, which start loud and die off over time, as if a word were bouncing from wall to wall in a wide canyon. By 1962, WQAM was using echo to highlight the position of a record on the station's playlist or survey. A typical sentence might sound like this: "Here's survey song number **FOUR** FOUR *four* four," with each succeeding"four" being slightly lower in volume.

Station Identification Jingles: The station ID jingles that were aired on Storz stations during the peak of the stations' popularity were made by a Dallas company called Production and Marketing Services, or PAMS. PAMS is generally credited with being the first jingle company to concentrate on producing jingles that identified and promoted radio stations rather than the products advertised on those stations.[1] Bill Meeks, president of PAMS in its Top 40 heyday, set out to make jingles that didn't just identify the station, but sold the merits of the station. He recognized that the more network radio programming dwindled in the face of television competition in the 1950s, the more radio became a *station* tuning medium rather than a *program* tuning medium. But listeners had trouble recalling which radio stations they listened to when queried by ratings services. Meeks felt there would be better recall, retention, and association with a singing station identification than with a spoken station identification. When Top 40 radio blossomed, jingles at first sold the new format, and then — as competitors arrived — sought to differentiate one station from another. A certain musical logo and a set of phrases about the station came to be considered not just part of the programming, but essentials of station promotion.

As PAMS productions evolved in the late 1950s, the company appropriated into its jingles the "signature" concept, in which the station's call letters were always sung the same way musically, so as to become the station's "signature." PAMS Sales Manager Jim West once referred to it as "a hit song in four bars." The first variation on the signature was the added phrase "Sincerely," as in "Sincerely, W-N-O-E," originally written into jingle lyrics by Bud Connell when he was programming WNOE against Storz's WTIX in New Orleans. When Connell moved to KXOK, he applied another letter-writing term, "Yours Truly," to the signature of a new PAMS jingle package. It became one of the best known and most memorable signatures in broadcasting, and was ultimately used on most of the Storz stations.

PAMS jingles were played so frequently and consistently on Storz stations that MacKenzie Repeater machines (explained in the next section) had to be employed to air them. The jingles "gave the stations a unique personality in the market," as Deane Johnson put it. He felt that "PAMS jingles were a major part of the programming thrust on Storz Broadcasting. They were very instrumental in creating the Storz sound. They were THE bright, uplifting element." Bud Connell countered that the PAMS jingles "were no more a major element than anything else — but the thing is, we used them *so often* that they *seemed* like they were more important." When Connell reprogrammed KXOK in 1961, he rotated PAMS jingles with a highly edited custom jingle series by Sandy & Green of Hollywood. The jingle package was based on the iconic song, "Meet Me in St. Louis" (with the "Louis" pronounced as "Louie"), a song originally written for the St. Louis World Fair in 1904 and revived by Judy Garland in the 1944 movie of the same name. If there was one popular song that every St. Louisan identified with, it was that one. KXOK calling itself "The Spirit of St. Louis" sealed the deal.

In the early 1960s, WABC, New York, negotiated 90-day exclusivity for PAMS jingles, meaning no other radio station could buy them for the first three months.[2] Having the jingles bought by an ABC-owned-and-operated station in the nation's largest market gave PAMS immense prestige. But Todd Storz had earlier negotiated an agreement that PAMS would not deal with any other stations in the markets where Storz had their jingles on the air. Deane Johnson remembers that in the late 1950s, prior to that agreement, Storz executives such as Bud Armstrong and National Program Director Bill Stewart would meet with PAMS president Bill Meeks in New Orleans, and would buy virtually everything PAMS offered — just to keep it off the other stations in Storz's markets. Bud Connell can attest that Storz's strategy worked. He recalls that in 1958 and 1959, he was buying jingles for WNOE in New Orleans from CRC and other Dallas-based companies, since his station was prevented from buying PAMS jingles because of their exclusive deal with Storz.

Storz program directors were able to write lyrics unique to their stations. When they were ready to produce new jingles, they would fly to Dallas to confer with PAMS officials in person. The meetings between PAMS and Storz executives lasted two or three days and PAMS picked up the "entertainment" tab. After 1961, the PAMS deals were handled mostly by phone after listening to a sample demo tape of the particular new series, although Connell and Grahame Richards (and sometimes Bud Armstrong) did meet with Bill Meeks at his Dallas headquarters.

In 1985, JAM Creative Productions (the successor to PAMS) released "The JAM Song" on a vinyl record that was mailed out as a promotional item. The record featured JAM's jingles as used by radio stations around the world. In effect, it was an audio "time capsule" of all that JAM had accomplished. You can hear "The JAM Song" at *www.jingles.com/jam/fans/jamsong.html* or go to *www.youtube.com/watch?v=1ZFubJmwXmQ&Feature=related* to see the mentioned radio stations' logos as the song plays.

Program Element Jingles: Jingles weren't only about the station's call letters and dial position. For instance, instead of concluding a newscast by merely saying, "Now here's the weather forecast," WQAM had a very upbeat jingle which sang, "Here's the latest Weather Word, from WONderful ... WQAM!" (Notice the alliteration of four Ws. It wasn't accidental.) Because Top 40 stations featured many more "specialized" program elements in an hour than did stations before or since that era, jingles touting even minor program features were plentiful.

Advertiser Jingles: If advertisers could be persuaded to buy a custom-made jingle as part of their message, that had the effect of making the station's overall sound more musical as compared to the "plain talk" heard on other stations' commercials. Pepper-Tanner Productions in Memphis was one of

Ten • Elements of the Storz Station "Sound" 145

several providers of budget custom advertiser jingles, sometimes offered on a trade-out basis. The radio station acquired the jingle at no cost or a reduced rate in exchange for a negotiated amount of air-time. Pepper-Tanner then placed commercials for national and regional advertisers in that committed time.

Commercial Production Music: Companies that supplied radio stations with production music offered recordings of brief instrumental *music bridges* (a single musical "phrase") that could be used as "separators" between program elements and as "punctuators" at the ends of others. They also offered longer recorded music *beds* for use under live or recorded promotions and commercials.

Rhyming: Disc jockeys who could make up rhymes as part of their chatter were in vogue for a while, and rhymes were a staple of commercial jingles, going all the way back to "Pepsi-Cola hits the spot, twelve full ounces, that's a lot." Rhyming was sometimes also a part of station ID jingles, although in subtle ways. WQAM had an end-of-newscast jingle that sang, "That's the latest, you've *heard* the *word*, on WON-derful, W-Q-A-M!"

Recorded "Drop-ins": *Drop-ins* (so named because the programming element was usually "dropped in" or inserted into the deejay's patter) were designed as surprise interruptions, thus demanding a response from the deejay — generally a humorous one. Sometimes the drop-in was funny in its own right, and did not require a response. KXOK deejays with alter-ego voices recorded their drop-ins before their shows, and responded to them on the air as if another person were in the studio. Mort Crowley, when he was KXOK's morning man, supposedly interviewed "President Lyndon Johnson" and other notable personalities with a slate of humorous questions and answers. Johnny Rabbit, as portrayed on KXOK by former child actor Don Pietromonaco, pre-recorded the voice of his alter-ego Bruno, and thus was able to sustain a five-hour-long nighttime show because the recorded Bruno filled half of the airtime with his funny comments or snide remarks. The teenagers in middle America loved it.

Syndicated Drop-ins came from many and varied sources, the most notable of which were contained in the "SuperFun" package featuring stars of the hit TV show *Laugh-In*, and written, produced and directed by O'Connor Productions in Hollywood. KXOK and other Storz stations used all of O'Connor's material for years. Dick Orkin of Chicago created *Chickenman*, a daily five-minute episodic feature which KXOK, WHB, and other Storz stations purchased and broadcast during the middle 1960s. The *Chickenman* of the title was very loosely modeled on Batman. For example, Chickenman was a shoe salesman on weekdays, so he was only able to drive around in his yellow crime-fighting car (the "chicken coupe") on weekends.

Contests and Promotions: These were unique in their day, and although they varied among Storz stations, they generated phenomenal levels of audience participation. In turn, that "proof of listenership" generated more advertising sales.

According to Bud Connell, there were two principle types of promotions: average-quarter-hour (AQH) enhancers, and cume (cumulative audience) boosters. The first were generally designed to elongate the tune-in time of the average listener. Typically, AQH enhancers were games with money or other prizes that required a listener to stay tuned for additional minutes, hours, or sometimes, days. An early example was KXOK's Money Mathematics, in which a new cash amount was announced every fifteen minutes. The listener's mission was to keep a running total, and when the on-air personality announced the end of that particular game, the first listener to call with the correct total won that amount. The listeners never knew how long a Money Math game would last, or when the deejay would call for a winner. Some games lasted for a few minutes, others ran for days. The bigger the jackpot grew, the greater the interest generated among the listeners — and providentially, the more that particular AQH promotion became a ratings booster.

"Cume boosters" were promotions that brought the call letters to the attention of people who were not normally listening. The idea was to convert those people into listeners, and thus increase the station's cumulative ("cume") audience.[3] Notable examples were Storz's original Lucky House Number, and Lucky License Number. If your house number, or your car's license plate number were broadcast on a Storz station — and you were not listening and missed collecting the cash prize — there was a good chance your friends or neighbors heard your name announced as a loser ... and they were likely to tell you about that too. Presto: a new listener was born! There were literally hundreds of such audience promotions originated by Storz personnel, many of which were copied by broadcasters nationwide.

Newscast sound effects: Newscasts on Storz stations were generally serious in content, but sought to be entertaining in their presentation. Early on, they were often underscored with the sound of a teletype printer (a noisy machine then ubiquitous in newsrooms), and they might be punctuated with the sound of Morse code dots and dashes or telegraph key sounds to separate stories and to highlight datelines. News stories were sometimes even separated by "transition" *music* that helped to set the tone. Special features could be incorporated into the newscast, such as KXOK's Exposé, an inside-Hollywood gossip feature, and Prophecy, a campy look into the future. Mobile news units gave live reports from the scenes of local events and offered highway traffic conditions during rush hours.

Mass Attendance Events: In the spirit of community involvement, Storz personnel had a finger — if not both hands — in many local events. Lou Cooley, KXOK's traffic reporter, raced giant boats in nationally-known Mississippi River events, and brought many trophies home to share with KXOK listeners. WTIX hosted *Beach Nights*, where over the years, hundreds of thousands attended free concerts emceed by the WTIX deejays. KXOK helped promote the June 1965 St. Louis appearance of Frank Sinatra's Rat Pack (Frank Sinatra, Dean Martin, Sammy Davis, Jr., et al.) for his favorite charity, Dismas House. The show featured Johnny Carson (the only time Carson ever sang on stage), Nancy Sinatra, the Step Brothers, Count Basie's Orchestra, and many more famous names. The star-studded show was broadcast closed-circuit to theaters in New York, Chicago, Los Angeles, and other major markets nationwide. In 1966, as recounted earlier, KXOK exclusively sponsored the Busch Stadium appearance of The Beatles, emceed by Johnny Rabbit, Nick Charles, and other KXOK deejays.

The Total Package: In summary, on a Storz station in the heyday of Top 40, there simply were far more "audio elements" in a given time frame than had ever been heard on radio before. The sound was fast-paced but usually without feeling breathless or rushed. The apparent speed of presentation was often more a product of program elements being presented very tightly packed together — that is, with no gap between one feature and the next. Storz programming also avoided the clockwork sameness that seemed to afflict most other major market radio stations. There was an overall sense of energy, excitement, and constant variety. Listeners often felt that if they tuned to another station, or turned the radio off, they would miss something that was interesting, important, or just plain fun.

Bud Connell asserts that there was a clear distinction between "context" and "content":

> The *context* in which the Storz sound was created included the format, the equipment, the sound-processing chain, the hit records, promotions, jingles, scheduled news, sportscasts, weather forecasts, and air personalities. The *content* included the attitude, the "fun" philosophy of the Storz stations, the station's image, events in the specific market, the clever writing-recording-execution of promos and commercials, carefully determined music selection, and the support and encouragement of personalities and their patter.

While Storz stations all adhered to most of the basic format elements, Deane Johnson recalls that each Storz operation reflected the style and taste of its local market. For example, WHB in Kansas City was adult and relatively conservative. KOMA in Oklahoma City was loud and aggressive. WTIX in New Orleans was a bit more rhythm 'n' blues-oriented with its music list.

Miami's WQAM celebrated both locale and climate by being "fun in the Florida sun." KXOK content reflected the rich history and culture of the St. Louis market.

The Storz sound was never "one size fits all (markets)." If today's nationwide satellite program distribution system had been available decades earlier, it is hard to imagine Storz would have used it to send one national program to all six stations. Even when Dick Clark offered a syndicated show that stations could air, the Storz stations opted-out in favor of local programming. Connell confirmed that most Storz personnel considered "canned" and nationally-fed programs to be inferior and obviously lacking local interest.

Unlike many Top 40 station operators, Storz Broadcasting never employed a programming consultant to coordinate the overall sound of the station group. Additionally, there was very little involvement from a national program director, except in the very early days when Bill Stewart ran the show. He was replaced by Grahame Richards after the 1959 Disc Jockey Convention fiasco. Deane Johnson recalled, "Grahame was certainly an ally, but he almost never commented on programming at KOMA. He helped me find jocks from time to time." Johnson concedes that in the early days, "Bill Stewart was a key architect, but by 1960, programming decisions were made by local program directors without much if any central influence."

Dick Fatherley observed that Storz radio did not have a "paper trail" culture as the McLendon stations and other Top 40 operators did:

> It was implicitly understood through the internal indoctrination of all of the key men in the programming ranks about what the programming fundamentals and the foundation are at a Storz station. Anything over and above that was at the discretion of the general manager or the program department.... The "old hands" in all of the stations knew what "the rules" were because they had all come out of Omaha. There was no paper trail, there was no [policy] book, there were no memoranda. However, the engineering department under Dale Moudy did issue a monthly memo.

3. Technical Production Facilities

For years, Storz's director of engineering, Dale Moudy, remained at the forefront of innovations in production and transmission equipment. Previous chapters have recounted how in the early days of Storz ownership of KOWH, Moudy developed a telephone answering machine that let kids talk to and hear from Santa. For KOWH and later Storz stations, he devised an hourly time-tone which was automatically triggered by a voltage sent by Western Union to reset the large clocks that were used in most radio stations of that era.

Moudy also equipped the stations with "limiting" amplifiers that — contrary to what the world "limiting" suggests — had the effect of keeping all elements of the program signal loud all the time. At KOWH, Moudy also developed an electronic "patch panel" which did away with the tangle of "patch" cords that connected various pieces of electronic equipment. Most of these innovations became standard practice at other Top 40 operations, and Moudy pioneered many of them.

In the mid–1950s, the typical Top 40 station's on-air studio was equipped with at least one microphone, an audio control console that allowed switching between sources and control of each source's volume level, and four turntables. Usually, three of the turntables were devoted to the playing of locally-recorded "acetate" discs, and the fourth was for the playback of hit music — typically from 45 rpm records. While the music was playing, the deejay (or less often, an engineer) could cue up three large acetate discs on which were recorded commercials, jingles, special features, public service announcements, etc. When the first disc of the three had aired and the second disc was playing, the first disc was removed and replaced with another one, and that procedure was repeated with the other turntables — often several times — until all of the scheduled elements had aired. Even commercials that did not require extra production (that is, they consisted of voice-only) were sometimes recorded anyway, to reduce the chance of errors requiring "make-goods"— the broadcasting equivalent of the "do-over."

If a station used acetate discs for all of the non-musical recorded program elements, when those discs began to sound scratchy from too many plays, they were discarded. The original announcement had probably been recorded on reel-to-reel tape, so it could just be "re-dubbed" to a new acetate disc. But if it was necessary to make a new disc anyway, why not re-do the announcement? It may be that the "wear factor" of acetate discs contributed to the freshness and novelty that was one of the hallmarks of some of the earlier Top 40 operations.

MacKenzie Repeaters and Tape Cartridges: Beginning in the mid–1950s, playback systems using loops of quarter-inch recording tape — the type used on reel-to-reel tape recorders — developed rapidly. In 1955, Louis G. MacKenzie began to offer his tape-loop "repeater" for sale.[4] The MacKenzie Repeater 500 model had five separate endless tape loops, any one of which could be started instantly. A silver-coated tag applied to the backing-side of the tape loop provided the cueing — when the tag passed a certain point inside the machine, the tape drive was disengaged, leaving the tape cued up for its next play.

These MacKenzie units were used by movie and television studios for sound effects that were played frequently, and they were employed at Disneyland to play noises at certain points during a theme park ride to heighten the sense of realism. Nationally, a few radio stations used them for short program elements that were repeated often. Deane Johnson recalled that Todd Storz saw an early MacKenzie Repeater, and ordered they be sent to all his stations. Johnson's assessment is that "they were *very* effective in the Storz programming." He was referring to their use in a live operation that required short, rapid-fire, frequently repeated program elements, which otherwise would have to be played from three or four specially recorded acetate discs, all of which had to be cued by hand. Instead, short, frequently used elements such as station identification jingles, laughs, applause, promotional one-liners, etc., were stored on MacKenzie Repeaters. See more about Mackenzie Repeaters and hear a frenetic newscast produced using one at *www.deanejohnson.net/audio/ KBOX_Action_Central_News.shtml.*

At KXOK, Bud Connell used the MacKenzie machines from the inception of Storz ownerhip, in both the deejay studios and in the newsroom. His *Action News Format (*which was dubbed *Essential News* at KXOK) was an uptown version of his *FUNdamental News* at WFUN in Miami, and featured approximately 37 sound cues in less than five minutes. All of the cues could be contained on the MacKenzie Repeater units. When the foreman of the IBEW (International Brotherhood of Electrical Workers) Local saw the cue sheet, he gladly relinquished operation of the machine to KXOK's newsmen who were members of AFTRA, the American Federation of Television and Radio Artists.

In 1956, the Fidelipac tape cartridge was introduced. Unlike the MacKenzie Repeaters, the recording tape was completely enclosed within the cartridge, and the plastic cartridge itself was designed for economical mass production. This cartridge's unique design eventually became the broadcast industry standard. In 1959, Fidelipac cartridges were used in the first engineered-for-broadcasting cartridge record-playback decks, which included a system for automatically re-cueing the cartridge to the beginning of the program element. That design also eventually became a broadcast standard. But during their first few years on the market, cartridge record/playback machines occasionally produced less reliable audio quality than transcribed discs, so some stations continued to use the transcription discs. In its Top 40 heyday, however, Storz Broadcasting was rarely a company that waited for second-generation developments.

Ironically, when easily erased, reusable tape cartridges that maintained their fidelity for long periods became the standard radio playback system, the result could be the airing of the same tired program elements for many months, or even years. That didn't happen at Storz stations, but did at less meticulous operations.

Program Automation Systems: In the 1950s and 1960s, radio programmers and station managers were interested in program automation systems for several good reasons: They offered a consistency in following a format that ego-driven and sometimes volatile deejays could not promise. They were a one-time capital purchase that could be written-off over time, whereas the salaries of air personalities were a continuing and often a growing expense. And the new machinery appeared to be "the wave of the future."

But reliability was a problem. "Cueing"—a signal from one machine that caused another machine to play—had to be absolutely reliable. Unfortunately, buildup of tape residue on the playback heads of early cartridge tape playback machines too often made that fail-safe intent more a hope than reality.

Given the known problems with automation equipment, it is odd that Storz considered automating any of his stations, since their formats placed heavy demands even on human operators. Deane Johnson believes that Storz made that decision based on either of two possibilities—"either the 'future view' that station automation was the coming thing, or 'economy'—one of those two things triggered it."

Dick Fatherley believed that WQAM, WDGY, and KOMA were probably automated as a money-saving move, but perhaps also in order to lead the FCC to believe that those Storz stations were more in control of program content, following the payola scandals. (Automation was not installed at WTIX, WHB, nor KXOK.)

In 1961, Storz made the decision to investigate program automation systems for his stations. Jack Sampson, who was then general manager of Storz's newest station, KXOK in St. Louis, remembers it this way: "Todd called me in 1961 to go to Reno, Nevada, to look at Paul Schafer's automation system—he owned a station in Reno. We did, and it was terrific. But Reno was a little market with not much competition. Schafer sold it to us as an economy deal, and not as a programming improvement. And we tried to make it an economy for us, and it didn't work—it didn't work for us in *any* way."

In 1963, WQAM's general manager Jack Sandler was interviewed for an article in the *Miami Herald* about his decision to *reduce* the station's reliance on automation. The reporter had visited the station in 1961, shortly after the system was installed, and remembered that back then, Sandler's face was "wreathed with smiles." Two years later Sandler said:

> We've cut back our automation.... It wasn't quite right for us. Not yet anyway.... Oh, it works perfectly. In fact, we're still using it for about 50 percent of our broadcast output. The trouble is that it has put us all under too much strain. A

couple of my people have actually broken down and cried. To keep it programmed has required more work from too few people. And another thing, the listeners have noticed the difference. They can detect the automation and they've complained about it. You know we worked very hard — and I think successfully — to create a definite WQAM image here and I think it's important to preserve that image.[5]

Bud Connell recalled a later incident that shook his confidence in automation when Storz was still toying with the idea:

In 1964, I was told to go to Minneapolis to look at the automation equipment at WDGY, which was then running automated, I think, in most of its day parts. I recall the date was some time in May because Bud Armstrong and I were playing gin rummy in one of our hotel rooms with an exterior entrance, and we didn't want anyone to see us, so we closed the drape in the morning; it was blue skies and a moderate temperature outside. When we opened the drape, there were several inches of snow on the ground. When we arrived at the station at the appointed time for a showing of their new equipment, it was operating flawlessly. Suddenly, a noise from the Control Room attracted our attention, and we saw a tape machine that seemed to have a mind of its own. A huge reel, which was one-and-a-half times larger than the hour reels we were used to dealing with, was throwing its tape up toward the ceiling and it was accumulating in the center of the room. By the time we walked in, the entire reel of tape was just a pile on the center of the Control Room floor ... and nobody said anything. We all just looked at it, and at each other, and that was the end of our meeting. I believe that was the end of automation for most of the Storz stations.

Jack Sampson summed up the introduction of program automation at the Storz stations this way:

I think automation was an experiment by Todd. He felt it was the coming thing, and he probably was right. But it was sold as a money saver, so we tried to do it that way. We automated KOMA, WDGY, and WQAM. It never did work right, and gradually each station converted back to live. The automation equipment now is much more sophisticated. At that particular time, it was awful.

4. *A Summary of What Appealed to Listeners*

All of the engineering advances, production techniques, and disc jockey creativity that were the hallmarks of the Storz station sound existed for one reason: to increase the *appeal* of listening to the radio, thus driving up TSL ("time spent listening," an audience ratings measurement), which, in turn, could produce greater revenues. Here is a list of some of the appeals that listeners were likely to associate with a Storz station:

Fun— The term "Fun Radio" was originally coined by Bud Connell for

WFUN, Miami — then competing with Storz's WQAM. However, it was adopted company-wide almost immediately when Connell brought it to St. Louis and installed it on KXOK. Almost everything else on the air was prosaic by comparison.

Stimulating— Listening to a Storz Top 40 station in its heyday gave you an eargasm.

Energetic— You had the sense that your favorite disc jockey was an athlete in his prime. He had great control of the timing and flow of the "game" — and he wanted to take it to you with a flourish and a smile on his face.

Loud— In fact, Storz stations *were* louder than their competitors — until the competition also installed signal processing equipment to "ride" the studio audio so that everything was equally loud, as well as maximizing the transmitter's modulation level.

Fast-paced— Storz disc jockeys weren't especially fast talkers — in fact, most spoke at a conversational rate, but they worked hard to fill every second of airtime with some kind of interesting sound. "Dead air" was not allowed. (The only major Top 40 disc jockey who ever got away with being droll and slow was WLS's Larry Lujack.)

Varied— On a Storz station, the listener was treated to many different audio elements in a short span of time. The sound was more "dense" than on any other station. There were more audio events broadcast during an hour than anywhere else on the dial.

Consistent and fresh at the same time— The two words are nearly opposites, but they coexisted on Storz's signals. The listener knew what to expect overall (there was format consistency), but the listener also enjoyed new elements that were being added regularly (freshening). That required a lot more work at the management and production levels, but being consistent AND fresh produced many delighted fans who kept their radios tuned to a Storz station.

• ELEVEN •

Four Sages at Four Stages

The definition of a sage is "a mature or venerable man of sound judgment; one distinguished for wisdom." The four broadcasters featured in this chapter all fit that description. Each provides perspective on important aspects of the development of the Mid-Continent Broadcasting Company/Storz Broadcasting Company over the span of more than two decades.

Virgil Sharpe offers glimpses of Top 40 programming precursors that were developed at KOWH in Omaha before the format had its iconic name.

Todd Storz, the man at the top of the company (but one not disposed to seek publicity), explains his radio programming and management philosophies in a rare public speech, delivered when the Mid-Continent Broadcasting Company was coming to national prominence.

Steve Labunski, the first of Storz's original station managers to leave the company, offers details of Todd's working relationships with his managers, underscoring Storz's belief that if the programming was right, sales would follow.

Finally, George W. "Bud" Armstrong, who was the longest serving member of Storz station management, provides insight into the procedures and philosophies that the company followed both while Todd Storz was alive, and following his untimely death.

I. Virgil Sharpe: Programming Precursors

[The following interview was conducted on October 13, 1965 — that date is listed on the first page of the transcript, which was supplied by Richard W. Fatherley. There are several sentences and phrases that have been crossed out with a pen or pencil. Words or phrases have also been inserted with a pen or pencil, in what looks like Fatherley's handwriting — but that is only a best guess. However, there is no reason to doubt that these are the words of KOWH's Virgil Sharpe recalling the early days of that first Storz station, and offering a unique perspective on Todd Storz's thinking and practices in his first few years as a broadcaster.]

TV began to make its inroads in Omaha late in 1949, and most of us out here in the Midlands of radio felt that if we closed our eyes it would go away. It didn't! We just didn't pay any attenton to or hear the cries of anguish of the people back east, as TV began to make more and more inroads into radio's income and take away more and more of its listeners. As a consequence, in 1950, the Hooper Ratings began to show a decline in radio listenership in our stations [in the Midwest]; and panic began to set in.

It was around this time that Todd bought KOWH. I first met Todd at a meeting of the Nebraska Broadcasters Association. Todd appeared at this meeting as the new owner of KOWH. He was a very thin, dark, young man — rather taciturn. My first feeling was: "What would you have to say that was of importance among those of us who have been in this business for some time?" However, as I listened to him, it suddenly occured to me that this young man DOES know what he is talking about and has done a great deal of thinking and knows something about the industry. I had a feeling that this would be someone who was in the industry for good and would make it his life's work — so watch out for him, because he's going somewhere!

Possibly he knew something I didn't. For example, we had a number of programs that we put together for the Nebraska Broadcasters Association. They were radio programs produced on discs. We sent them to all member stations to play. Todd's observation was: "Why don't you make tapes instead of sending out new discs which cannot be replaced and which, after all, are expensive? Why don't we put this material on tapes, send them out and let them use the tape and send it back; erase it and send it back out again?" [Tapes were new at this time]. In this way, you could use the tape a number of times. It struck me as being so practical that I asked myself "Why didn't I think of that?" Many times during the years I was with Todd, I found myself saying "Why didn't I think of that?" This is some insight into how the young man's mind worked. He was constantly considering the problems of radio and consequently, on a moment's notice, could pull out an answer, or at least a pretty good answer for most any problem. Todd seemed to have the facility of shutting out of his considerations, a lot of outside extraneous experiences and problems when it came to radio. I have a feeling, from my own conversations with him, that Todd pretty much ate and slept and lived radio. Because of his technical knowledge of broadcasting, he looked at radio from a complete viewpoint: as a person who understood radio from the time the spoken word went into the mike [sic] until it came out of the antenna — and indeed beyond that. He had the complete picture of the function of a radio station.

Todd and I had various telephone conversations concerning problems that were mutual to all radio stations. TV was cutting in, and radio, in general, felt that it was on very, very, shaky ground. As a matter of fact, many times we felt a terrific earthquake going on beneath us. I can remember the summer of 1951 when I called Todd one morning and told him it had begun to look like I was going to have to move out of the operation I was in, because things were not progressing satisfactorily. He said, "Why don't we have lunch?" Well, we did, and that time he said, "Why not come with me?" in his way, which is difficult to describe. He said quietly, "I can't pay you what you're worth, Virge, but perhaps later on we will be able to get your salary to a point where it will be what you are worth to the organization." I accepted that, because this was the way I had always worked in the past. I have some of my own peculiarities, one of which is — I want you, and an employer, to pay me what I think I'm worth and what you think I'm worth. If I'm not producing, then let's talk and do something else. I came up in that way and accepted Todd's offer.

Now, I have a suspicion, and that is this: Before Todd hired anyone in those early days, he took it upon himself to find out a great deal about them. I don't think Todd ever went into any kind of negotiations, whether it was for buying a radio station, employing someone for an important position with the company or even moving them up, or even perhaps for any kind of discussion with anyone in the radio industry; unless he had found out considerable about this person. He was always interested in knowing what appealed to this person. What line of reasoning did he take? But more important, Todd wanted to know, and usually found out, how he could influence this person. This he did in a number of ways. I know he felt that a person who has a large vocabulary, usually equates with a person of intelligence, who regularly uses the vocabulary. I know also, for example, that he spent a great deal of time with the dictionary, improving this very aspect of his character; so that he understood words and knew how to use them. And he used them! He wanted to be able to express himself as well as he could. In general, to express himself better than the people with whom he was working and the people with whom he associated. This he would do very quietly. He understood that people accept you at the face value you place on yourself and this, I think, added to the stature of the man. This was without a great deal of grandeur and personal feeling, and without any great egocentric appearance. This he merely did quietly; and he was able to gather unto himself, if you will, erudition, dignity, power and control; and that is why I don't think there is any question that the man began to develop from those early days on. I think he began to see and know the things he needed to do for himself. This is where a great deal of credit is due him. It's a picture of someone becoming important and successful in the field of radio, and so he neglected no opportunity in any area to do anything, or any kind of research, or work that would add to his ability to get ahead in the radio industry. Todd felt that in order to get where he wanted, to control radio properties, to be able to persuade other people, and to become an authority in the field, he must not neglect any area of self-improvement.

When I joined KOWH, I came in with additional positions — some air work, sales and programming. I didn't really have a title. For a period of seven or eight months, I did some air work. Yet, Todd and I conferred quite frequently and we began at that time to make some changes in programming. Then I filled a gap created when Todd's assistant left the organization. I moved into that position and it was about this time that Todd began to look at the programming.

One of the all-music programs on KOWH seemed to be getting a higher rating than some of the other programs during daylight hours. In addition, an air personality carrying the program seemed to pick up and lift the ratings as well. At the same time Todd bought a survey. The research was conducted by an administrator of a local [Omaha] university, who was in the business of making surveys at various times. He conducted — more or less on the side — an industrial psychological testing bureau. He knew a lot of people, had many ideas and had surveyed a lot of areas. He was also interested, I am sure, in what people listen to, and why. What motivates them to listen? What he was after was: What do men listen to and what do women listen to? The survey revealed that music had a high appeal to most everyone. It was a question-and-answer thing. The results of the survey were then compiled, with the end result being that *music* was an important item and was an essential ingredient in the motivation of listenership in radio. So, whenever we programmed blocks of music, the interest would go up. However, being independent, KOWH had a few "canned" programs, one of which was called Kitchen Klatter. It was piped to us from a station in Iowa and featured two very nice ladies talking about problems, horticulture, flowers, recipes, etc. [Kitchen Klatter had gone on the air over KFNF in Shenandoah, Iowa, in 1926, and eventually became the longest-running homemaker program in the history of radio.] I think this was the tipoff to us. We discovered that the ratings went down at this time, and it took almost two hours after this show for the audience to tune back again. Todd and I reviewed the ratings, and I said, with some trepidation, "I think we are going to have to drop this Kitchen Klatter program, because the personality programs and the music appear to make sense to the public." Being ten years older than Todd, I would find myself approaching that which was different from what we had been doing, or that on which we might disagree, with a feeling of trepidation. It might be the trepidation that exists when you bring up something with the boss in any case, but I'm not sure. I think I was concerned with what he thought of my intellectual processes. This was the kind of effect he had on you. I think you would be concerned with "He wanted to know without even saying so." [*sic*] In a few simple words, he would ask, "How did you arrive at this decision?" or "Do you think you have the answer?" or, "Have you thought about this?" or "How long have you been toying around with this?" He wanted to know what you thought and how you arrived at a conclusion. To those who knew him, he did it quietly and with a few words. He was one of the sharpest, most concise minds I have ever known. It could get to the heart of the matter or could clear up or could find out, could search out the answer in the fewest possible words. Nothing ever kept him from getting to the point. His mind arrived at it quickly — in a flash! He said "Do you realize how much money this brings to us, this Kitchen Klatter program?" (We received it and fed it to some other stations.) I said, "Yes, I know. I know how much money it is bringing, but I think in time we are going to get it back and more too." He had a habit of sitting back — he would place his hands together and sort of rock just a little bit; and he would turn and gaze off and look toward the end of the room. Then he turned back again and said "You really think so, huh?" And I said "Yes, Todd, I do." I didn't have any idea what was going to happen. The following Monday it was off the air, and the preceding personality that was on, continued for another hour.

Todd had another facet to his character, which I want to state as correctly as I can, and that is this: Todd interpreted the successful operation of a radio station in terms of revenue. And I mean that exactly as I say it, without adding anything to it or detracting anything from it, or reading anything into the meaning. A radio station, to him, was a successful operation if that station was bringing in as much revenue as it possibly could. He felt, quite objectively, than any kind of a business should be interpreted in terms of what it produces and how well it produces. And he said this to me many times. As a result of this feeling, he also had a theory: We discussed this theory many times. It was, if your station is sold out, your rates are too low. This happened on KOWH, and as a result we had at least four rate increases while I was there. So you see, he had an interpretation concerning the operation of a radio station as a business. Todd felt that in a country such as ours, which is based on the capitalistic system, that a business was supposed to produce revenue and therefore, if it produces revenue to the best of its ability, it is successful. In radio, these two things go hand-in-hand. You've got to have the audience, or if you will, the circulation, in order to get the rates, to get the business. But he was working with a radio station, back in the beginning, where we had a segment of time (Kitchen Klatter) that was producing revenue. Therefore, there isn't any question that Todd had problems facing him. And so he was no doubt thinking along the same lines I was about the money and maybe on the *other* hand wanted someone's judgement to fortify his own — to make the move. From then on, there was never any question in Todd's mind as to what kind of program produced audiences.

The problem now was to refine the idea, to get personalities who had appeal and to draw attention to KOWH. Now this was another one of his ideas about radio. And to a certain extent mine, which I've always had. A "radio station" occupies a certain niche in the community. It needs attention drawn to it. It has to have that intangible *something* that causes people to tune in. It has to have these things in addition to the actual programming you have on the station. Sure they used to tune in to listen to Sandy Jackson, to listen to Jim O'Neill — of course they did. But they also tuned to KOWH — because KOWH had Sandy Jackson and Jim O'Neill and some of those other people that we had on the air over the years — they didn't sound alike, but the station had an overall "SOUND" (which I think is neglected today in radio) which I think is important. Todd would go to great lengths to get attention drawn to the radio station. I don't know of any broadcasters who would have had the guts, at that time, to put one of his disc jockeys up a tree in one of the parks, throwing away dollar bills and then broadcast bulletins about a man who had suddenly gone berserk and was out in the park throwing away dollar bills, with a huge crowd gathered around the man and being arrested and put in jail. People by the thousands showed up to bail him out. All of this caused a great deal of attention. But this is what KOWH needed! We had what we thought was a good product, but we had to draw attention to it first, just like the old story of hitting the mule in the head with a 2-by-4 — first you've got to get his attention! Todd knew this. Either instinctively, or he had found it out through his own experience, but he knew it. I don't need to go into a lot of the other promotions, because we had them galore and they are a matter of record that can be picked up, I am sure, from some of the other people. Our Treasure Hunts, our Lucky House Number, all of these things. Todd also had one other theory, which I think is important: What do you give away and what is important to people in terms of prizes, or the things they would like to have? Other radio stations many times had contests on the air where they gave away appliances and most anything you might think of. Todd thought people were always interested in money. So when you have a prize, make it money and make it a good amount of money.

So, Todd gave away money. You see, Todd had thought this out very thoroughly. What is the common denominator of need? MONEY! And so therefore, he was absolutely right in terms of appeal. You could appeal to almost everyone with money, whereas merchandise and that sort of thing, while it has its appeal, is not the same thing. But, when we did go to a prize of merchandise of any kind, we went big. It would be an automobile, and this, of course, was also a common denominator. I don't think there is any question that the Todd Storz operation was the first radio station in the country to offer a $100,000 prize for finding a hidden treasure. We tied up all the traffic in Council Bluffs and I don't think they have recovered from it yet! I don't know how many cars were over there. Only the fertile mind of Todd Storz could come up with a way to handle the idea. I can remember when we broached this treasure hunt. There were three or four of us sitting in the office talking about it and we said, "Well, now what about going big — about $100,000." I can't attribute any of these things to one person, because they were off the top-of-the-head "skull sessions." Todd would listen and occasionally throw in a word to direct the area of discussion and keep away from spending too much time. So we made a decision — let's have a *big* treasure hunt. We had had before that, smaller treasure hunts where we would go out to one of the parks and we would hide checks all over the place, for $10, $50, $100. And then we would give clues over the air. Hundreds and hundreds of people in cars would arrive and go through this treasure hunt and find the $500 check in a capsule hidden in a mud puddle or in the bottom of a lake somewhere. All of which is very colorful and added to the general personality of the station. But the biggest treasure hunt of all was the $100,000 one. But who except someone with the mind of Todd would say at the end [of the planning session], "How would we protect ourselves?" We now, of course, could afford to give away $100,000; but it would make a terrific dent in the overall worth of the radio property. What will we do? And we sat for a moment and then he said (and I remember this so plainly) with a twinkle that occasionally appeared in his eyes, "What about Lloyds of London?" Well, there is your answer. We insured the $100,000 prize with the local representative of Lloyds of London! Lloyds arranged that they would hide the checks. The did so on dark nights — one of the most "cloak-and-dagger" situations imaginable! The policy was signed (I think it cost $4,000) and all the details of hiding and the clues were given to us. We didn't know where the treasure was hidden. This was, of course, an absolute necessity. The clues were good. I remember the last day of the treasure hunt. The cash prize went down as the clues got closer. I believe the final amount was $1,000 and it was found! Interestingly enough, one person kicked the tin can in which it was hidden during the hunt, when it was still [worth] $100,000! Nobody will ever know how close they came.

This is just one of the things I mean when you talk about how Todd's mind worked. It would never have occurred to me to think of someone insuring the treasure hunt. I understand that the man who did this had a rather lucrative business later on, handling this promotion for other radio stations — of insuring treasure hunts. Things happened after Todd did something. They always did. If it was done on KOWH, then it would appear somewhere else.

I imagine that after we began to roll and our ratings began to come up to the 40s, 50s and 60s in Hoopers, that probably no radio station ever had the number of visiting "firemen" sitting in hotel rooms with tape recorders, putting down what we were doing at KOWH in the early days. By "firemen" I mean men who ran the radio stations across the country, who would come in and air-check us. Later, as a matter of fact, one of the recording companies here in Omaha took orders regularly from other stations to tape everything we were doing.

In the beginning we said, let's try to pick music that is modern, and I don't mean in the currently modern sense. Let's also stick with the old favorites. Let's use judgment in picking out our music. So we got the various personalities together and said, "Now you pick out your music and build yourself a show every day around this music and let's see how it sounds." And it began to work. We did this way back in 1950. Some of this was being done when I came there. [But] Jack Sandler had a sports half hour in the late afternoon. He had previous to that been sports announcer on the old KOAD-FM.

The repetition of call letters, the time, the temperature grew. We would have discussions every day. What is it that people want that we are not giving them? This was Todd's question over and over again. What is it that they want or will listen to that we are not giving them? And he would ask this in many forms, in many ways, not only of us, but of himself. And this, I think, kept us moving. And so what began to evolve is this: Number one, what do people want to know? They want to know the time. So let's not get very far away from letting them know what time it is. Number two, it was understood in general, but not nearly as much, the people want to know what the temperature is. So let's put the time and temperature together, so they know what it is. Number three, they want to know what is going on in the world, but they don't want to know it all the time. And they don't want it carried out at great lengths. So let's keep them informed, but let's not keep them overly informed so that their attention wanes. Our five minute newscast idea began to come into being. It didn't constitute too much of an interruption. However, Todd was concerned about news as a music interruption. This was one of the things he and I argued about. The broadcast day evolved into five minutes of news and fifty-five minutes of personality and music. We clicked, not by accident, but by [the] choice of James O'Neill, who was our afternoon man. He began to play the quite popular music, the late music that was coming out. And he began to carry a good deal of what appealed to the younger people and suddenly the rating on that began to jump. Any place we saw a rating jump, we figured if it jumped there, it would somewhere else [on the broadcast schedule]. And this is how, first of all, the top 10 on our station originated. We played it first in the afternoon and then we moved it into all the other segments. Then we put popular music around it, but we made sure that the top 10 was repeated in every program segment. And the ratings began to climb. The top 10 began to move in, in 1952. And by the time Todd had negotiated for WHB, we had the format! We began to think of it as a format and as a whole programming operation. Top 40 then was another development, but came about 1956. We were concerned with the top 10 and a little more, but again some latitude was given to the personality. Later in 1956 and '57, as we got a program director on a corporate level, the Top 40 *must play* situation evolved. Really, the repetition of the 40 tunes occurred around 1956 in the Storz organization. But you must also realize that some other things were happening. *All* these things began to evolve, because imitation is the sincerest form of flattery, and imitation began to arrive. It didn't take long for record companies and other radio stations to realize the impact of the repetition of popular music. "Let's send out and put in public places, the Top 40 songs for this week or next week." "Let's have people vote on it — let's have housewives vote on it." There became all kinds of options on this thing. You see, Todd always started something.

We used to have twelve spots an hour, not counting short breaks spots. The interesting thing was, this (when we were filled to the hilt with spots) did not decrease our popularity; but it seemed on the other hand, that people liked to listen to a station with commercials — in spite of all the things that were said by the competitors. Todd and I together (and I say this because we spent one whole day discussing it) arrived at a conclusion which I still think is true. That you can talk all you want to about young listeners and the fact that this kind of programming has appeal to teenage listeners; but we had a feeling (and I still have a feeling today) that the youthful mind runs from about age twelve until about thirty-five — in his likes and dislikes. The reason for this is, that there were two or three occasions where we would have a couple of the jukebox operators make a determination as to what was played most on their jukeboxes in the areas where young people didn't go (or at least were not supposed to go); and that would be some of the bars that had jukeboxes. We found adult tastes to be very close to the type of music that was supposed to be listened to only by teenagers. So we arrived at the decision the people in the movie industry arrived at long ago — that if you are going to appeal to popular demand (and this is in no way derogatory to the American public), the tunes have to be simple, they have to have a beat, they have to be recognizable, and they have to be in an area that is appreciated by the so-called young group. We felt that most Americans are pretty young at heart. And I think that is where the Storz appeal really hit the mark and became really successful. The appeal to people who are young at heart — who are interested — who are forceful — who are active — and who use radio as kind of an outlet of their own for their emotions and ideas. Todd knew this.

In conclusion, let me paint a picture of Todd's personality. More so than perhaps many people who worked with him realized then or will ever realize, Todd was very sensitive. He was not sensitive about what people said about him or about the situations of radio stations or the business or about his own personal attitudes. He hesitated and disliked very much to bring hurt on anyone. He carried — to the outward eye — all the aspects of the cold, steel-like, steel-nerved executive. This Todd was *not-inside*. He was tremendously interested and would become vitally concerned about everything that was done. But it wouldn't show too much. When it was necessary sometimes to get rid of someone, while he could discharge the person or fire him, this was something he would much rather have the manager of the station do. Todd was manager and I was program director or assistant manager. But then when I became manager and vice president, he moved into the other stations and overall operations, and he moved out of that. There was a time when Todd was still the executive officer of KOWH.

When discussing another person, he would merely say "It seems to me that this particular salesman we have had for quite a while, and we have been more than kind to him — we have given him a great deal of time to see whether he is going to produce, and I just think that for the best of all of us and for his own good, that it is going to be necessary that we get rid of him. Doesn't it seem that way to you?" And then I would usually agree, because he would probably be saying something I had thought of myself. I must say that in the operation of KOWH, there were only a couple of instances when that situation arose, because generally it would be the other way around. I have enough experience in the business to catch this before it got that far. Invariably — with maybe one exception — he would say, "Well, what do you want to do?" And I would say, "Well, I would like to do so and so." In general, you always found yourself with the burden of proof on you. He had the knack of making you state your reasons why. He wanted to hear what you wanted to do, because he was interested in how people's minds worked. He spoke rather quietly, in a low, well-modulated voice. He had a distinctive voice.

Todd was a sensitive man. He would, many times, spend some sleepless nights thinking about things he had to do to people and the reasons for them. I know also that as time went by and the company grew larger and it was necessary for him to be in the top executive position, I am sure from what he told me that he missed very much the personal contact within the group. He enjoyed a great deal the first stages of KOWH, where we were all struggling for success. Once it grew, he knew what to do.

One other thing about him — and he said this just before we acquired WHB — "It seems to me that there isn't any more difficulty in running a large station in a large market than there is a small station in a small market. They have exactly the same problems, you do exactly the same things. It is only that you are talking to more people with more power. Therefore, the future of this company lies in good stations, on a good wavelength, in large markets."

II: Todd Storz: The Man at the Top During the Advent of Top 40

This section is based on photocopies of a talk given by Todd Storz at the University of Georgia. The title of the talk was "Independent, Alive and Healthy." The photocopies were obtained by Dave MacFarland from the McLendon Stations archive in Dallas, Texas, in 1971. On the front page is a notation by Gordon B McLendon to his secretary: "Billie — mimeo copies — 1 for each station + several for my desk. GBM."

Eleven • Four Sages at Four Stages 163

The date when this speech was given does not appear on the document. The talk itself refers only to programming on "the station," which would have been KOWH. However, the photocopy of the talk lists two of Mid-Continent Broadcasting Company's other stations: WTIX and WHB. The latter station was bought in March of 1954. It does not list WDGY which was purchased in December of 1955, so it is likely the talk was given in the intervening period. But it could also be a photocopy from 1955 of a talk that had been given some years earlier.

Because the speech begins immediately with references to stations and sales, and since at one point in discussing KOWH's contests Storz says, "All the other things I have mentioned today you are perfectly free to adopt if you choose," it is probable that the talk was given to an audience of fellow radio broadcasters. The talk covers a multitude of factors in the evolving radio landscape: changes in audiences, choosing music, presenting the news, specialty programs, unique promotions, and more.

> In so many stations today, principal effort and thought is directed almost entirely toward sales. Our philosophy that audience comes first seems to be almost unique. Yet, in almost any other business or industry, their product certainly comes first. The manufacturer of a new soap product, for instance, certainly would not devote a lot of money and effort toward sales until he was convinced that he had a product of high appeal and comparable or better than his competitor's similar product. Audience and sales are not always truly compatible. Sometimes it is necessary to sacrifice sales at least for the moment, to take the long term approach to programming, product and audience.
>
> In order to do its best toward audience, a station must retain complete control of all its programming. For example, if a station has built a highly successful block program, let's say from 12 to 2:00 PM, and a preacher should present himself at the station with an offer to purchase from 1 to 1:15 PM, cash in hand, the station manager's plight is obvious. If he accepts the program, he knows he will hurt his 12 to 2:00 PM block program. If he turns down the program, he knows he will be sacrificing immediate revenue. Our answer without hesitation would be "no" to any program that didn't fit our overall program schedule. We can justify the loss of immediate revenue by the firm knowledge that we will have that revenue many times over, over a period of time, by adhering to proper program standards. No matter how good a station's audience is, I do not mean to suggest that the sales department can be retired. Advertisers will never be knocking down the door to buy time no matter how successful your operation. Nevertheless, with proper programming and audience, the sales resistance is greatly lessened. This is particularly true on renewals, since with a large enough audience, results are virtually assured, and, after all, results are what advertisers buy.
>
> We do not believe that our mission in this world is to educate people, because radio is a purely voluntary listening habit — that is, the listener is free to turn the dial or turn the set off— programming cannot be based on compulsive listening. We feel that a station's public service value is closely parallel to the station's rating.

For that reason, our programming is all directed to mass listening. We omit virtually all types of minority programming even though, in some cases, the minority may be large. Of necessity, a large part of any independent station's programming consists of music. Therefore, the station's music policy must be given a great deal of thought. We play only popular music. No hillbilly, no religious, no classical.

Admittedly, some of these minority groups are quite large; but, even though an individual's favorite choice of music might be classical for example, we are sure that he would still enjoy some popular music. It's the common meeting ground of all music today. Within this broad category of so-called "popular music" we have narrowed our field even more by placing particular emphasis on the so-called "hit" current pop tunes. These are not played to the complete exclusion of all other types of popular music, but our entire music format is built around these tunes as a basis. Since all stations have almost exactly the same music available to them, it would seem at first thought that all stations would be reduced to a common denominator. So, almost everything else that we do is our particular way of presenting our programs which are primarily popular music. Only by complete attention to the many details in proper relation to the basic music program is it possible for one station to have ten or even one hundred times as much audience as his competitor who has exactly the same music available to him.

All the rest of our operations might be termed "Showcasing the music." Radio has long been known as an excellent medium for news. We use one newscast each hour, "five minutes before the hour." We feel sure that our listeners want news, and although they have shown signs of crossing us up in the past, I think we now have a common meeting ground with them on our news. At the beginning of the Korean War, the rating on all of our news showed a substantial increase over the program preceding and following the news. But in early 1951 this trend was completely reversed and we realized that something was wrong with our news policy. Realizing that we could only find out from the listeners, we decided to call approximately 100 of our listeners who had recently sent mail of one sort or another to the station. Several of us made these calls and had long discussions with each person called, whenever possible. While admittedly 100 people represent a very small sample for any survey, the response we obtained was unanimous, and we thought the sample was truly significant. Almost without exception, the listeners asked us to stop all news completely.

They didn't like our news — they only endured it to get to our next music program. Summing up their comments, we found that their interest in news was very low, practically negligible, except for a reasonable amount of interest in local news and a great deal of interest in what I will kindly refer to as human interest stories, more particularly Hollywood divorce scandals and the like.

Confronted with this information, we gave serious thought to discontinuing our news, but for many reasons, not the least of which was the fact that our news was very successful in a commercial sense, we decided to make one last try at salvaging our news and making it interesting to the listener. Two additional wire services were ordered, which gave us the facilities of all three of the major wires — AP, UP, and INS. Newscasters stopped their "beats" and instead were asked to spend all available time preparing each newscast by using, virtually without rewrite, wire stories. Knowing what the listener wanted, we gave it to them in large doses. The truly important news happenings of the day were summed up in a 30-second spot bulletin-type summary of the important national and international happenings. The rest of each 5-minute newscast was devoted to the local news, the sensational news they wanted, and weather, which was also of considerable interest. Needless to say, there was a great deal of reluctance on the part of the newscasters in following this policy. However, we felt that if we had continued our policy of straight news, we would soon have few news listeners at all. As it is at present, we have very high news listening and occasionally, we can sneak in a truly informative news story and our listeners have listened to it before they realize it.

On-the-scene reporting of local news seemed to us to be a good path to pursue. We constructed a mobile unit which had a complete short-wave installation, making it possible for us to broadcast from any spot in the area on short notice. These broadcasts are principally of disaster- or casualty-type stories such as plane crashes, drownings, fires. etc. We interrupt our regular programs to broadcast these direct stories.

THE DETAILS

(1) DO SOMETHING. In retrospect, I know we have made many mistakes, but we have never regretted them. Without action, a station soon decays. Doing anything at all is really difficult. It's an old saying, and very true, that *a few people make things happen, many watch them happen, and the majority have no idea what has happened.* With enough action, the time will come when listeners will be afraid not to listen to your station because they might miss something.

(2) PERSONALITIES. All of our programming has been built around disc jockey personalities. We feel that this has helped the station receive a true personality of its own and has a considerable appeal to our listeners who are principally housewives. Many stations shy away from personality programming because of personnel problems. I would be the first to admit that personality-type programming does produce a good many additional problems, but we feel that the value of this type of programming outweighs its disadvantages.

(3) SMOOTH PRODUCTION. Even though most listeners are not too critical, we believe that they are quick to realize sloppy production when they hear it, even though their realization may be almost without a true knowledge of what caused their dissatisfaction. Smooth production can be achieved only by constant attention to many production problems. No dead air, elimination of discs with high surface noise, good cueing, and a general fast-moving pace are only a few of the things that must be considered.

(4) STUDIO APPEARANCE. It's true that today most stations do not entertain in their studios many people from the listening audience, nor do many station advertisers or potential clients ever visit in the station. Nevertheless, we believe that a good operation has to grow from the inside out. I am not advocating an elaborate studio layout, but I think it's extremely important to have a place that is neat and clean, modern, and in general, a pleasant place to work. I know how difficult it is to justify a large expenditure in remodeling and redecorating studios.

Intangible as it may be in its immediate effect on station revenue, we are sure that money spent in this way comes back over the years.

(5) GIMMICKS. A lot of stations have written to us asking for a list of gimmicks and special promotions we have used. We have never kept an actual list or count of these things but because of the interest shown, I am going to give you a few of the ideas we have used. They may not all be of interest to you. But they do constitute an important part of the action I talked about earlier, and cumulatively I think they are responsible to a large extent for our success.

(a) LUCKY HOUSE NUMBER. We started this contest in 1949. All the other things I have mentioned today you are perfectly free to adopt if you choose. This particular contest we have under copyright and it is syndicated to a number of stations in the U.S. and Canada for a moderate franchise fee. We have found it to be highly successful and so have many other stations. Each and every listener has one chance to win. Their winning number is their house address. We obtain this address on the air by spinning a bingo mixer to get the numerals, and a standard 16-inch transcription [disc] to get the street name or number. The transcription contains the name and number of every street in the city. It was originally transcribed at 33 ⅓ rpm. On the playback, we start it at 78 and then stop the motor. The name the transcription stops on is the street name used. A building jackpot is used, and each time there isn't a winner, the jackpot increases. This contest has great appeal because everyone knows his or her street number. Also, since many of the street addresses obtained at random are nonexistent, the jackpot tends to build up to a high value before a winner is found. We have had cash prizes as high at $3,000.

(b) MYSTERY VOICE. This is an adaptation of an old contest format. Listeners wishing to be a "Mystery Voice" send in postcards. The announcers call a listener and her voice goes over the air as she repeats a limerick after him. Any of her friends who may be listening can call in and if she is identified as the "Mystery Voice," she and the person identifying her share the jackpot.

(c) FLYING SAUCERS. During the recent high interest in flying saucers, we transcribed a 1-minute announcement which was repeated over the station several times for several days. An echo chamber was used on the voice and it went something like this: "This is KOWH in Omaha, calling all flying saucers. Recent reports have indicated that flying saucers are cruising in the vicinity of earth. If any operators of flying saucers should wish to establish contact with earth, please call us on 20,000 kilocycles — our transmitter engineers are now monitoring 20,000 kilocycles for calls from any flying saucers. This is KOWH, Omaha, Nebraska, U.S.A., Earth, calling any flying saucers. Please come in on 20,000 kilocycles." After each of these broadcasts, the actual monitoring of 20,000 kilocycles was rebroadcast over KOWH; however, I am sorry to report it consisted only of a few static crashes and pops — no flying saucers that time.

(d) CHARITY. It's difficult to classify charity as a gimmick. Certainly it isn't. But our charity program is part of our overall action and I believe has contributed to the station personality. We maintain a charity fund. Some of the money is donated to this fund by the station. The balance of it comes from our listeners. We use it in many different ways. Usually, it's to help out someone after a true calamity when other charity isn't available to them. For example, a small oil refinery had an explosion in which two of the workers were killed. The employer went to the widows of these workers and gave them approximately $12 each, which was the amount of salary the men had coming. The widows and their children were destitute. In this instance, we gave them each $500 to help cover burial expenses.

During the recent polio epidemic, local hospitals were unable to give proper treatment to some polio patients because of the shortage of all-wool blankets needed for wrapping hot packs. Standard appeals on the radio and in the newspapers for all-wool blankets had failed to produce a sufficient number. The situation was really critical. We interrupted our programs and went on the air with a request that listeners call the station to donate blankets. After a few calls were in the station, the announcer called the mobile unit and repeated over the air the name and address of each person wishing to donate an all-wool blanket. The mobile unit started making these stops and at the first few stops interviewed the housewives on the air. Before the day was over, we had a backlog of 3,000 donations and had enlisted the aid of 30 trucks and drivers donated by local businessmen to pick up the large number of blankets. By the third day after the first broadcast, all hospitals treating polio patients had enough blankets on hand.

(e) OMAHA AFTER DARK. Many of you are familiar with the format we used on this broadcast so I won't go into great detail. *Time* magazine carried a pretty complete story on the broadcast. In this particular case, almost one year of preparation went into a broadcast that took only 55 minutes of air time. A German recording machine constructed along the lines of a fine watch was used to collect the data for this broadcast. Our announcer concealed the machine on his person. A fine wire ran from the machine to a dummy wrist watch which was actually a microphone. Armed with this ingenious setup, he began to frequent illegal places in the city, principally those places violating the gambling and liquor laws. Our listeners heard actual broadcasts from inside those places, together with incriminating evidence. Places were named. Names were named. This particular broadcast probably attracted more attention than anything we have ever done.

(f) TREASURE HUNT. Listeners were told that we were going to have a treasure hunt. If they wanted to participate, they were to send a stamped, self-addressed envelope. In this envelope we returned to them complete instructions and a banner which was to be placed in the back window of their car. We knew that on the day preceding our actual treasure hunt we had put out 18,000 banners. The question in our minds was, how many of those people would actually participate. The treasure hunt took place at noon on Sunday. Our first broadcast said: "Calling all treasure hunt cars. Here are your first instructions — assemble in the downtown Omaha area." From this moment on, the rest of the day was pretty much of a nightmare. It was later established that the cars, if lined bumper-to-bumper, would have stretched over 60 miles. Moments after the first broadcast, our local law enforcement officers had succeeded in finding me and informed me I was being held for helping create a riot and that we must discontinue the hunt. However, I felt that since we had promised our listeners a treasure hunt, we should go ahead with it. As a result of this position on my part, I spent the rest of our treasure hunt time securely locked in our local jail. However, from what I hear, I understand that it was a good success.

Listeners were given clues which led them to a dummy check. The check could be redeemed for cash at the station the next morning. There were five checks with total prizes of $1,000.

(g) [no title]. In 1952, Omaha was besieged by a very threatening flood of the Missouri River. Throughout the entire flood crisis we carried many programs of direct broadcasts from the field which were transcribed and put in our files. On the first anniversary of the flood in 1953, we thought it would be of interest to our listeners to hear some of the old flood broadcasts. The program ran for 55 minutes and was identified at the beginning and end as being a rebroadcast from the flood of the previous year. Nevertheless, by the end of the broadcast, 200 volunteers had appeared at City Hall for flood duty, and over 100 families had evacuated low-lying areas although the river level stood at below normal.

[CONCLUSION]. A great deal has been said recently about radio having reached maturity. Sometimes I think radio has become too mature. Many stations seem to pride themselves on doing nothing, each day's broadcast being just as passe' as the last day's. When people stop talking about you or radio in general, something is wrong. Although today we have over 2,500 radio stations in the U.S., whereas at the beginning of the war there were less than 800, and although television has become another competitive factor in both audience and dollars, I look for the future to hold new rewards for those stations willing to do a little something extra to attract and keep that fickle cornerstone called audience. Without it, none of us could survive. With more of it, our future will be bright.

III. Steve Labunski—The Top 40 on Networks and Major Market Stations

In 1957, several of Storz's key men were recruited by other broadcast interests, including American Broadcasting Network (ABC), the Katz station representative company, and competing Top 40 station operators. The first Storz manager to leave would be Steve Labunski. Labunski had been an effective salesman at "The New WHB" and was a major contributor to its spectacular billings growth and sales volume. He became a vice president and the general manager of WDGY immediately following its purchase in 1955. That station's history of small audiences, little revenue, messy bookkeeping, a weak sales department, and a complicated and expensive antenna system were all liabilities which Storz had assumed. But Labunski had proved fully capable of helping Todd and Bud Armstrong fix WDGY's problems.

In an interview with Richard W. Fatherley in 1965, Labunski provided several glimpses of Todd Storz's guidance:

> I could change a lot of things; local promotions, this, that, and the other. But, I couldn't change the music policy even a little without permission ... his [Storz's] theory was — and he followed it religiously — if the product is good

enough people will listen to the station. If they listen in large enough numbers, then the rest of the ball game is obvious.

There's an audience and it's measured. You have rating points. Then you have a sales story for advertisers. This was considered to be a normal consequence to the product and its promotion.

Todd was the first to recognize there are morning-men with huge audiences who do everything wrong "by the book" but, somehow people listen to them. The disc jockey who thought of his audience in their terms — the audiences' terms — was the man Todd was looking for. Todd loved the air personality who reached for the radio and patted the "housewife" on the fanny. It was that kind of touch Todd was looking for.

Once a month, Storz called each of his general managers. I would talk to him in an organized, formal way, once a month on the telephone when our financial statement came out. We would talk about expenses. He would needle me about them if they went up. In going over the statement he was obviously making himself do a necessary management function. He could ask good questions and needle you about things you'd forgotten. You had a copy and he had a copy. It would take five to ten minutes. He would say, "Ugh huh ... yeah ... ugh huh," and give you a lot of skepticism. You'd say, 'I think we'll do better in this area,' and he'd say, "Ugh huh ... yeah ... didn't you tell me that a month ago?" It was a kind of gentle cross-examination. But the time you'd spend on the phone in anguishing re-appraisal was when the ratings came out! That was far more important — or the resignation of a good disc-jockey. If I lost the best salesman — fired him — he would say, 'Ugh huh ... yeah yeah.... I assume he'll be replaced?" But, if the morning-man quit me, this was reason for an hour's conference on the thought. How good was he? What would happen to the audience? The point is, Todd was more concerned about the loss in programming — any reversal in programming — than he was about a temporary sales problem. He was program-oriented. He still believed, through it all, that the P & L [Profit & Loss] balance sheet was going to take care of itself if we kept the programming good, and therefore the rating high.

He felt about programming the way a singer feels about his voice. He doesn't care whether he drives to the concert in a Cadillac, whether he has a Chesterfield coat on, or something else. He cares whether his voice works. A bubble-dancer is worried about her complexion, not whether her gut aches, but whether it shows. Right? Todd Storz was interested only in the essential things.

As a human being, an individual, Storz was remarkably indifferent about a great number of things, one of which was that a great number of things didn't interest him at all. I could mention peoples' hobbies, fields of endeavor — he had a fairly one-track mind. He was interested in radio, in the competition, the challenge; the excitement of taking a station from ninth place to first place, quadrupling the billing, and seeing the competition scratch their heads trying to figure out what happened

He loved the pioneering phases of it. When things were going relatively well, in kind of the second phase of the Storz company when things were running themselves, or managers were running them autonomously — and he was consistent: if he let you run it, he let you run it — it gave him less to do. Some of the fun was taken out of it, since the experiences had been lived through once or twice. I think the fun went out of it for him at this point.

My only observation of this was in a general way. All of the stations, more or less, were on a very strong footing. Not all in first place on all surveys, or all without problems; but in general, the pioneering phase had ended, and as soon as it had ended Todd began to lose some of the spirit that motivated him before.

I think that maybe down deep he realized that there were others in his company who had to go elsewhere; that there was a way of outgrowing even this company of his, and the temptations outside, for which he'd helped prepare us, were simply too large to keep us all there.

I don't think he took it personally. We remained friends, but I was somewhat out of touch with him. I always thanked him for things, and when I ran a station in New York [WMCA] that got into first place, I sent him a telegram saying, "Thanks for showing me how so many years ago."

I was unashamed at my love for him as a person, and my respect for him as an operator and a boss. I think, perhaps, that I held a little special role in that I was the first management guy to leave Storz. I wasn't the last, but I was the first. If he took it a little hard, he never let on, and I never knew it.

My admiration for him is unreserved. I think he was a giant contributor to radio broadcasting when radio was prepared to fall flat on its ass. It was prepared to do so, and I think Storz was one of a handful of people who kept it from doing that.

Steve Labunski was recruited by Robert Eastman, president of the American Broadcasting Network (formerly ABC radio), to install its *Live and Lively* format on ABC's existing radio network. *Live and Lively* was not a nationally distributed disc jockey program. Rather, it featured live performances by musicians, singers and performers. In an interview with *Minneapolis Morning Tribune* columnist Will James, Labunski claimed, "ABC resembles WDGY eighteen months ago." ABC radio, he said, had been compromised while corporate attention was given to building ABC television. A split had been negotiated, with the re-formatted radio network being "separated from the TV network so it can fight on its own."

In late September 1957, Labunski appeared with Dallas-based broadcaster Gordon McLendon at a luncheon meeting of the Radio and Television Executives Society in New York City. *Broadcasting-Telecasting* described the Labunski-McLendon showdown as "spirited discussion." Labunski said, "The very thing which Gordon McLendon and Todd Storz and others are doing to help assure local radio a future are substantially the same things which American Broadcasting is doing to help assure radio networks a future. Therefore, if Gordon McLendon has a future, so do we." McLendon retorted, "The sole function served by radio networks ... is to provide coverage of national and international events. In every other area of programming, local radio and/or television is superior."

After just five months of *Live and Lively* network programming, ABC abandoned the idea as too costly. WHB business manager Roy Lollar said Storz referred to it as "The Great Experiment." The experiment had failed.

Labunski resigned from ABC saying there was little left for him to do. He subsequently was hired as general manager of WMCA in New York City. There he installed a "polite" (i.e., not "loud and crazy") Top 40 format and recruited Herb Oscar Anderson from WDGY to host the morning show. Labunski later was named president of NBC radio and became one of America's top network management figures. But his 1957 departure from the Mid-Continent Broadcasting Company is important not only because he was the first of Storz's managers to leave, but also because he tried to help two major U.S. radio networks remain relevant against competition from independent Top 40 stations — and television.

Another important Storz pioneer followed in Labunski's path: Todd's boyhood friend Dale Moudy had risen to be Storz's vice president for engineering, and was a key player in the development of the Storz group's "sound"— especially in a technical sense. He thus became a valuable asset to ABC Radio after joining Labunski at that network as a "special services" consultant to ABC's owned-and-operated radio stations — such as powerhouse WLS in Chicago, which adopted and polished the Top 40 format. "The motivation was money, a lot more money," Moudy admitted.

IV. "Bud" Armstrong — Storz's "Right-Hand Man"

George W. "Bud" Armstrong was Todd Storz's "right-hand man" beginning with the earliest days of KOWH in Omaha. When in September of 1953 Storz bought his second station — WTIX in New Orleans — he named Armstrong as vice president and general manager of that facility. Under Armstrong's guidance, the low-powered 250-watt signal at 1450 AM went to the top of the New Orleans daytime Hooper ratings, beating powerhouse network affiliates like 50,000-watt WWL. WTIX even managed to be a close number two at night. It was a remarkable achievement, since WTIX's coverage radius was—at best—about 25 miles. It was Storz and Armstrong together who countered New Orleans' WDSU 1280's *Top 20 on 1280* countdown record show with the WTIX *Top 40 on 1450*. WTIX's countdown show eclipsed WDSU's program by beginning one hour earlier and ending one hour later, and by offering twice as many hits. On WTIX, the current-hits format that had been evolving since it was first aired in 1951 on KOWH got its name — *The Top 40*—which became so common that it is now a standard dictionary term.

When Storz purchased the powerful regional signal of WHB on 710 kHz in 1954, he promptly sent Armstrong to Kansas City as general manager so that Armstrong could install the Top 40 countdown show that WTIX had pioneered. Rumor had it that Storz promised Armstrong a lifetime job upon making WHB successful — which is what happened. Armstrong's protege Fred Berthelson, who had helped WTIX make New Orleans radio history when the station enjoyed a 3,000 percent boost in advertising revenues, succeeded Armstrong as WTIX's general manager.

By 1955, "The New WHB," under Armstrong's leadership, had captured a near-50 percent share of the Kansas City radio audience, according to the C.E. Hooper ratings service. Perhaps more important, WHB endured as Kansas City's audience-dominant station until FM listenership began to displace AM audiences nationwide.

Bud Armstrong was the first Storz general manager to be named a vice president of the Mid-Continent Broadcasting Company. In response to that promotion, Harold Soderlund — who had been one of Storz's early mentors — wrote with tongue-in-cheek: "Dear Vice-President, Congratulations in getting to such an elevated position, dear boy." But, in fact, Armstrong was the right choice for the lifetime job with Storz radio that Todd had promised him. His programming know-how, get-the-job-done attitude, and interpersonal business savvy would have put Armstrong at the top of Storz's "A-List" anyway. He would go on to be named executive vice-president of all Storz operations in 1958, and that same year was elected to the boards of both the National Association of Broadcasters and the Radio Advertising Bureau.

When Storz died suddenly in 1964, Bud Armstrong continued to steer the Storz Broadcasting Company with a steady hand. Storz received a lot of press coverage while he lived, because he was head of the company, the one who chose the stations to buy, the one who oversaw the retrofits and additions as the Storz "formula" was installed. And it was Todd Storz who risked his own money — not his father's. But it was Armstrong who ran the stations day-to-day. Lots of work; not much glory.

Metaphorically, if Todd Storz was a comet that glowed brightly as it traversed the sky for an all too brief time, Bud Armstrong was the constant North Star who made dependable navigation possible long after the comet was gone. His importance to the Storz radio enterprise from its Omaha debut all the way through the disposal of the entire Storz station portfolio in the mid–1980s cannot be overestimated.

Born in 1927, George William "Bud" Armstrong attended public schools in Chicago and Omaha. He completed his secondary education at Creighton Prep School, graduating in 1945. He served in the navy from 1945 through

1946, then got his Bachelor of Science degree from Creighton University in 1950. He had done a little part-time radio work during high school at KORN in Fremont, Nebraska, and then during his college years in Omaha, at KOWH. Armstrong began as an announcer, but shifted over to KOWH's sales staff in 1949 when Todd Storz bought the station.

Even in those early years, Bud Armstrong had the look and the behavior of the executive vice president he would become. At 28 years of age, dressed for business in a suit, white shirt with a tab-collar, a regimental tie, and shined shoes, he was the farthest thing from the gum-chewing, toe-tapping young guy in penny loafer shoes who was trying desperately to be "cool." In business company, Armstrong spoke in a matter-of-fact voice using few words, and in a tone that was low and slow. When relaxing with Storz colleagues, he often smoked an expensive cigar. "Todd hated the smell of it," recalled Bud Connell. (Storz smoked Chesterfield cigarettes, a frequent advertiser on his radio stations.) Like the hood ornament on an expensive automobile, "Bud" Armstrong's cigar became a symbol of his authority to his subordinates. He was given to late hours playing gin rummy with colleagues while sipping a Tanqueray Gibson on the rocks, and chatting about business. His employees knew they had a good job. In Armstrong's company, they felt they had new importance.

Armstrong was responsible for installing the Storz "sound" at each of the new stations the company acquired. WHB newsman Charles Gray affirmed, "It was done by Armstrong. He was the man who brought in the plan and executed it. Todd Storz was in and out, but Armstrong was the man who was on the scene. He was in charge of—as we called it—'The Army of Occupation.'" Gray added that the engineers suffered a bit of "culture shock" when Storz took over a station because "Armstrong's rule was, when you worked in a Storz station, even as an engineer, you worked in a coat and tie."

On August 2, 1971, Dave MacFarland recorded more than an hour of conversation with Armstrong at the Storz Broadcasting Company home office in Kiewit Plaza, Omaha, Nebraska. In the interview, Armstrong makes a strong case that hard work, rather than flashes of genius, are what made the Storz stations successful. Portions of that interview are produced below.

> ARMSTRONG: Broadcasting was an avocation as far as I was concerned. I was going to the university, and getting into law school, so all I intended to do was to practice law. Broadcasting was an avocation, a development during my high school years ... I worked briefly for a station during one summer while I was in high school, and I kind of enjoyed it, so when I got out of the service and came back to go to college, I got a job with a station here, while I was going to college. And it so happened that the station was purchased. And Todd Storz was a friend, and it also happened that they bought a station here. And I

stayed with it during the finishing-up of college, and then Todd convinced me that maybe I ought to stay in the radio business.

MACFARLAND: You were an air personality?

ARMSTRONG: Oh, yeah, I had done some of that — some sports announcing, and some air work. I was probably the world's worst disc jockey, so at least, if I ever learned anything about programming, it was how *not* to do it, rather than how to do it. And I did some sports announcing — sports was also an avocation. I was very avid about that. So I've done play-by-play and color [commentary] in several sports: football, baseball, hockey, and basketball.... KOWH was kind of a hodge-podge-programmed thing, and we did many different types of experimentation on the station before evolving down to the format. I think the first step in the right direction was getting me off the air! [laughter] ... independent stations, in those days, had to be resourceful on their own, because they couldn't get the networks, and so they had to think up something else.... We had block-programmed country and western. We also had — and this was indigenous to this market, because of the population makeup — we even had an hour of polka programming at one time. We didn't originate that idea because it was already popular on another station, and we were just trying to take it from them ... this would be 1949.... And this kind of thing continued for the first year. But, we also, at the very outset of the thing, had the idea of block programming popular music, which we did in the afternoon. These other experimentations were pretty much done in the morning. And they gradually, over a period of months, pretty well disappeared, so that by the middle of 1950, we pretty much had *a* consistency — it wasn't the final end product, but it was a consistency in popular music and news.... We thought that somehow the functions of networking would pass to television from radio, and they *were* at that time. And we also realized that in terms of, for example, news coverage, the independent station that was unable to affiliate had to do something a little different in news from the stations that were affiliated, and that they — because of their networks — could do a better job of national and international news. We concentrated on local news ... [in regard to air personalities, we wanted people who] whatever they said would be relevant, or would provide a service feature, but would not bore them [the audience].... We wanted air personalities who could be creative, occasionally humorous, informative, companionable — all those things, in less than 30 seconds. And some people say that's a new thing. But that's just an old basic that we had years and years ago. And our disc jockeys had to be completely inoffensive in terms of — well, for example, one "blue" remark and the guy is fired. That's from day one, and it's true today. I don't care. Life's too short. We don't feel that our role is given to that sort of thing, and so we simply won't do it. We think our programming — and the demographics show it — appeals to a broad segment of the audience, so we are not going to get down in the filth museum, and we're not going to get into the double entendres, and we're not going to get our people to be so "in" that they're out. And as far as I'm concerned, if a guy gets that "in" on the air, he is *out*, and I mean now — he won't even finish his show. It's just that simple....

You were asking what are some of the things that characterize our operation. Well, one of them is that we *do* have standards, and we *do* have policies that extend into all these areas, that try to prevent any kind of slipshod operation, where anybody can take advantage of a situation. We're just not free and easy with these things, whether it's music or commercials. We're very fussy about copy we'll take. There are things that even the [National Association of Broadcasters] Code approves that we will turn down. Because, we aren't that interested in making the last buck, that we would jeopardize our audience.... And the reason is that, fundamentally, we are an "operating" company.... When we went into the business, we went in because we believed in the radio business, and believed in it not simply because it looked like a way to make a fast buck. I think a lot of guys got into it that way. We're operators. We love the thing itself. Todd did very much, and I do too. And this pervades the whole company. We didn't get in to buy a station, pump it up, sell it, buy another one, pump it up, and sell it. In the whole 22 years that we've been in operation, we have sold exactly one radio station. And that was the one in Omaha [KOWH].

MACFARLAND: Of course, you bought stations that were already in very good markets.

ARMSTRONG: That's right. And not only that, but we never changed a call letter, ever. So, we're really not a gimmicky operator. And we never have been. Our basis is to establish ourselves in a market with local people as much as possible, local management as much as possible, and to become a part of the community...

MACFARLAND: Would you say that at some of your stations there is more personality or chatter on the air than at others?

ARMSTRONG: I suppose it will vary some.

MACFARLAND: What would be the determining factor? What seems to succeed?

ARMSTRONG: ...I don't think there are any great differences in people from one city to another, but there *are* nuances of difference. And there are established personalities in some markets that *still* are personalities. Oddly enough, there are guys who have been very, very well established — and I mean personalities in their own right ... and that station could do almost *any*thing and they'll hold the audience. But, that same personality, if he leaves that station and goes across the street — zero. They can't take it with them. The station may not be able to do any wrong while that guy is there, and they can violate their own format. They can talk too much or talk too little or play all the wrong records or run too many commercials or whatever, and the guy will hold the audience on that station. Now, there aren't very many cases of this left anymore, mostly because all of these personalities who were indigenous to stations like that, did start believing their own publicity and took a better deal on the Coast or whatever, and had to find out the hard way that they can't take it with them. So most of them have jumped the traces and gone, and where *are* they today? They just aren't there. So, there seems to be a marriage of certain stations and certain guys — that *he* could do everything wrong and *they* could do everything wrong, and they'd still be number one. But, you take him off the air, they can do everything right and they're hurt; and he can do everything right at another station and he can't get a listener.

MACFARLAND: So, it's "individual cases."

ARMSTRONG: If all these answers were as simple as a lot of people would like to make them, it would be very easy to operate a business.

MACFARLAND: Well, I hope you'll forgive me for asking the questions anyway.

ARMSTRONG: Perfectly all right. But, unlike some of our associates, I don't feel we necessarily have all the answers all of the time. And I think that probably is the secret of our success, that we don't have anybody who has written "the word".... The geniuses, and the guys with a patent-medicine approach to everything — they come and they go. But your major broadcast entities, which are reasonably run, for maximum effect on the integrity of the listener as well as appeal to the listener, and on integrity of the company from a business standpoint as well as programming interests ... they go on, and they'll be here long after the fast-buck artists are gone, the fad artists are gone, the guys that have one book and that's the only way to run it. It's too bad that programming in radio has had so many people who thought you could just write a book, a format, and live by it like the Ten Commandments. They'd be just like horses with blinkers on: they just couldn't see that the application isn't totally universal. They're doctrinaire. There's only one way to look at things, and everything else is — well, this simply isn't true...

When I was first starting, and I was at KOWH and later at WTIX and the early days of WHB, I was sometimes told by advertisers about our teenage music, which ... turned out to be ten years later what the conservative, middle-of-the-road station was playing as *his* format. So, these things evolve. And we do have a lot of wonderful little old ladies in tennis shoes that listen to our stations [laughter] because they like the personalities, or they got used to it, or they just want to spend a little time and keep a little younger. Sure, we have a generally little younger profile, but not bubblegum and not teenybop and not.... We don't drive our programming home for the kids and then figure everybody else is going to listen. That's an old theory: you drive for the kids and then whatever else you can cadge on the side.... That leads to too many pitfalls, I think. That leads to maybe doing some things that we wouldn't want to do...

I think that in terms of people, or programming, or station operation, that there is no contradiction-in-terms between solidity and "keeping up with it." And, that which is most creative and most innovative and which will last the longest is that which is carefully thought out, carefully applied with objectivity, logic, and common sense, rather than with hunch, with impulse, with so-called intuition. The latter things generally tending to characterize those who feel the "free-spirit" type approach — and that's a very board oversimplification. You know the old saying: sin in haste, repent at leisure. And I think most things, even if they appear to be spontaneous, and "with it,' if they are really good, they probably have been very well thought-out and very well planned ... [or] programmed out in a man's mind. This business is too big and too important to be flying by the seat of its pants anymore.... A lot of these guys have always wanted to do it with mirrors. Well, you can't do it with mirrors anymore; you can't do it with flashes of brilliance and genius and waking up in the middle of the night with the greatest idea in the world — that's just nonsense. And there are still practitioners of this art. And people who have fooled a coterie of guys: Oh boy, he's a genius, he's really smart, he really know where it's at, he can really pick out the records. I don't know whether they feel it by osmosis or a guru speaks to them in a foreign language or what, but it really is all nonsense.

Because, sure, anybody can have a good idea, and w're always looking for good ideas. But the good idea generally starts out as a germ of something, and then is expanded by a lot of people who add their flavor to it, add their bit.

MACFARLAND: It's not just one man's genius.

ARMSTRONG: No, there is no such thing. And there never was. Anybody can have an idea, from the general manager, or president of the company, right down to the receptionist. This is perfectly normal. And it isn't the province — any more than a guy can predict hit records, or predict how an audience is going to react — it just simply is not true. And people are not fooled by that. You start with the germ of an idea which can originate anyplace, and you work it around and mold it, sculpt it, test it against other things, and you use rational and logical means to do these things — you keep your head about it all the time. Then you come up with a finished product.

MACFARLAND: You take time and you get help

ARMSTRONG: That's right. You take time and you get help. That's a good way of putting it ... the circus-performer-type people — most of them are already out and the rest of them are having hard times and ultimately will be out. Because, for a lot of them, when it becomes a sophisticated thing — and this is true for a lot of other people too — when it become a sophisticated thing, when there is an encumbering sense of responsibility that one cannot succeed without having, they suddenly lose interest, because they don't want to be bored with all that kind of thing.... We are not in that kind of business. We have senses of responsibility. Responsibility comes before doing your own thing. Our responsibility is to our audience. You *cannot* operate a business by the seat of your pants and be responsible. It's just that simple. And most of the genius types, or the guys who can forecast the future, or who get quoted a lot in the trade press — they come and they go. Like the bands.

By the mid–1960s, at the height of Top 40's popularity, the Storz stations under Armstrong's management were producing more than $20 million annually in billings.

• TWELVE •

The Decline, Sale and Legacy of Storz Broadcasting

The death of the 40-year-old radio "genius" Todd Storz on April 13, 1964, stunned the broadcasting industry, many notables around the nation, and of course his colleagues and friends.

A wire service report datelined "Miami Beach" began with these three sentences:

TODD STORZ — HEAD OF STORZ BROADCASTING COMPANY, ONE OF THE NATION'S LEADING INDEPENDENT RADIO GROUPS — WAS FOUND DEAD AT HIS HOME IN MIAMI BEACH THIS MORNING. HE WAS 38 YEARS OLD.

THE CAUSE OF DEATH COULD NOT BE DETERMINED IMMEDIATELY BUT A STORZ SPOKESMAN SAID THE YOUNG BROADCAST EXECUTIVE APPARENTLY DIED AS A RESULT OF A CEREBRAL VASCULAR OCCLUSION — A STROKE.

STORZ LIVED ON EXCLUSIVE SUNSET ISLAND NUMBER 3 WHERE BRIDGE GUARDS MAINTAIN PRIVACY.

Brief newspaper quotations of the coroner's report on the cause of Storz's death were corroborated by Richard W. Fatherley, who obtained a copy of the full autopsy report in 2009. The first two autopsy findings were "1. Pulmonary congestion and edema. 2. Marked coronary and aortic narrowing, minimal to moderate atherosclerosis." The detailed report included this sentence: "The sections of the coronary vessels show multiple focal and segmental areas of marked atherosclerotic narrowing with areas of 90 percent occlusion." There was nothing in the report pointing to foul play, or that Storz had taken his own life.

However, an autopsy finding from the Office of the Medical Examiner,

Twelve • The Legacy of Storz Broadcasting

Dade County, suggests that narrowing of Storz's coronary blood vessels may have been but one of two factors. A second possible cause of death might have been barbiturate intoxication. Under "Toxicological Findings," the heart blood is described as "3.15 mg percent barbiturate." Under "Gastric Content (barbiturate identification)" the finding was "Same Rf as Tuinal." Tuinal is a barbiturate that was sometimes prescribed to insomniacs to help induce sleep. It can produce the same "high" as alcohol abuse. Today, Tuinal is known to be dangerous because the amount of the drug that induces drowsiness is only slightly less than the amount that causes death, and tolerance to the drug can be built very quickly.

Storz may have been taking Tuinal because his left adrenal gland had a rare (and perhaps previously undetected) pheochromocytoma. The onset of such tumors is most common from early adulthood to middle age. Todd was two years shy of turning 40 when he died. A pheochromocytoma tumor causes the release of too much epinephrine and norepinephrine — two hormones which control blood pressure, heart rate, and overall metabolism. In Storz's

---

FORM ME-5

OFFICE OF THE MEDICAL EXAMINER, Dade County   AUTOPSY FINDING

NAME OF DECEASED   TODD STORZ   4/13/64   3:50 p.m   CASE No. 64-815

CAUSE OF DEATH:

    Pulmonary congestion and edema due to occlusive coronary arteriosclerosis associated with hypertension(Pheochromocytoma of left adrenal) and barbiturate intoxication.

TOXICOLOGICAL FINDINGS:

    Cerebrospinal fluid: negative for alcohol

    Heart Blood : 3.15 mg% barbiturate

    Gastric Content: 111 mg.total

    Gastric Content(barbiturate identification): Same Rf as Tuinal

Robert L. Catherman, M.D.
ROBERT L. CATHERMAN, M. D.
Assistant Medical Examiner

RLC:jkb

Autopsy finding from the Office of the Medical Examiner, Dade County, Florida. This photocopy suggests that the narrowing of Todd Storz's coronary arteries may have been only one possible cause of his death. Accidental overdose of the barbiturate Tuinal may have been another factor. The medical examiner's report leaves the exact cause of Storz's death "up in the air."

case, all three would have been too high. His symptoms might have included palpitations, severe headache, weight loss, and sleeping difficulty. That last symptom might have prompted Storz to take slightly too much Tuinal. But by pointing out both the "areas of 90 percent occlusion" in his coronary vessels, and the "3.15 ng percent barbiturate" in Storz's heart blood, the medical examiner left the exact cause of death up in the air.

Bud Connell says, "It was common knowledge in the company that Todd used medicine to relieve migraine-type headaches, and to sleep. His oversensitivity to the medicine may have contributed to his death, but taking too much intentionally was just not possible. He wasn't the type."

On the day that Todd Storz died, Connell issued a memo addressed to "The Entire KXOK Staff." Small portions have been omitted below for clarity.

> The departing of Chet Thomas [KXOK's "legacy" station manager, who would be replaced by Jack Sampson] and untimely death of Todd Storz within such a short span of time has made us perhaps receptive and vulnerable to rumors.... Controlling interest (60 percent) of the Storz Broadcasting Company is still held by Robert Storz, Sr. He remains as Chairman of the Board. No doubt Todd Storz's 40 percent will be held in trust for his three young children until they become of age.
>
> Todd Storz was a brilliant man and had adequately and carefully laid out the future of this chain. There will be no changes in our mode of operation or personnel...
>
> There is a great future for each of us, and this company will continue to be the top radio chain in the country, as it has been since the early '50s. — Bud

It is sad to have to say so, but Connell's final sentence stating "this company will continue to be the top radio chain in the country" was one of the few instances when Connell's foresight failed him. At the time, he had no way of knowing how different things would be with Todd's father running the company.

The next day, Jack Sampson received a telegram at KXOK from Robert H. Storz, asking Jack to be a pallbearer at Todd's funeral, to be held at Trinity Cathedral, Omaha's largest Episcopal church, on Thursday, April 16, 1964.

With most attendees still shocked by his too-early passing, Todd Storz's funeral was an especially somber affair. The eulogies were plentiful and elegant, and the pallbearers were Storz's own key management executives: Bud Armstrong from WHB, Jack Sandler from WQAM, Fred Berthelson from WTIX, Dick Harris from WDGY, Phil Trammell from KOMA, and Jack Sampson from KXOK. Many well-known people and dignitaries packed the church, as did the radio industry's elite, such as fellow Top 40 station owner Gordon McLendon and his father B. R McLendon, Harold Krelstein of Plough (another Top 40 group owner), station representative Robert Eastman, and executives of the Blair station representative company. Most also attended the lavish reception that followed — a tribute to Todd and the company that so reflected his style.

Jack Sampson remembers that day:

> There were some wonderful eulogies. What they said about him made us realize how fabulous he was. Although we loved him and believed in him, we were so close we didn't see all that — but he really was a hell of a guy. Todd's death was a personal blow to all of us. His gentle, but intelligent ways made him a joy as an employer. After the service, Robert H. Storz had a reception for the former staff members of KOWH and the out-of-town dignitaries, which read like the *Who's Who* of radio at the time.

Two months later, on June 30, 1964, Robert H. Storz sent a letter addressed to "My Storz Broadcasting Company Colleagues." It was on Storz Broadcasting Company letterhead, with a return address of 767 Forty-First St., Miami Beach, 40, Florida — the new headquarters that Todd had recently built in a warmer, sunnier climate — in both the meteorological and emotional senses of the words. It read:

> Since the Storz Broadcasting Company has grown to become the dynamic leader of our industry, it is to be expected that our competitors will try to take advantage of the untimely death of Todd Storz. We have been advised that rumors are being circulated that some or all of the Storz Stations are to be disposed of ... these rumors are not true. The Storz Stations are not for sale nor do we contemplate a change in ownership or management. Be assured I have not discussed the sale of any of the Storz Stations with any individual or representative.
>
> Just fifteen years ago this month Todd Storz and I began the Storz Broadcasting Company. It has been my good fortune to have been closely associated with Todd in building and expanding the Storz Broadcasting Company since the date of its inception...

In June of 1966 — two years and two months after Todd's death — Robert H. Storz sent the following memo from the Omaha office to Jack Sampson: "Dear Jack, In spite of our handicap of operating the Home Office from three locations, the loss of Todd and Jack Sandler, the Storz Broadcasting Company will complete a record year on June 30, 1966, which is the end of our fiscal year...." A bonus check to Jack was enclosed, which Robert noted "represents the maximum amount the government permits for profit sharing." It was back to business.

With Todd gone, his vision of how compelling radio could be needed to be carried on by the men and women who had worked for him. They maintained his programming principles for as long as they were allowed to do so. Unfortunately, that time span was short. As former Storz manager Deane Johnson put it, "Todd's death [and the subsequent control of the radio stations falling to Todd's father] brought about a shift from a 'programming company' to a 'money company.'" That is a succinct description of the crucial difference in focus that transpired because of Todd's passing.

Jack Sampson said that during his years as a Storz station manager under Todd Storz, "It was kind of the philosophy of 'Build it and they will come.' We built the programming, and the sales came. If the manager had any brains at all as far as sales were concerned, the sales came. But it was on the strength of the programming." Deane Johnson concurred:

> Yeah, I think programming was on a throne. I sensed when I was there, certainly, that programming was dominant; that program directors were thought of more highly than sales managers in an operational sense.... One of the curiosities when I went with Storz was in *Broadcasting Yearbook*. At almost every station in America, the personnel were listed as the general manager, sales manager, program director. Under Storz stations, it was listed as general manager, program director, sales manager. I think it made a statement that Storz was a programming company.

Sampson recalled, "A lot of times in my stations — and particularly at KXOK — a lot of times we didn't even have a sales manager. I did it. And I think a lot of the Storz managers did. I carried an advertiser list myself and I did the sales. Hell, I didn't want anybody else to do it, because I knew what I wanted, and I wanted to teach them how to do it." Left unsaid by Sampson was that he was delighted to let Bud Connell (his program director and later, operations manager at KXOK) literally "run the show" prior to Todd's death. Connell said "I think that program directors had free reign in Kansas City and St. Louis, but the other stations had *some* control from the home office." That was about to change.

When Robert H. Storz took over the company following his son's funeral, he soon discovered that the most successful members of the management team operated their individual stations like personal domains. Often, when the elder Storz attempted to exert control from his home office in Omaha, he encountered resistance and a litany of good reasons to continue activities the way they were. But Robert's nature, plus his long experience in banking and finance, could not permit any operational policy to continue that was not "pyramid-style" and directed from the top down.

On one of the elder Storz's initial visits to St. Louis, Jack Sampson and Bud Connell entertained him at The Colony Lounge, a hotel bar in St. Louis's trendy Clayton suburb. When Jack excused himself from the table to make a phone call, Bud Connell seized the opportunity to bring Robert up to speed on current trends in the broadcast industry, and to perhaps usher in the next necessary phase of the Storz Broadcasting Company. At the time, Connell was unaware that a decade-and-a-half earlier, Robert and his son Todd had experienced a money losing venture with the Omaha FM station they had bought along with profitable KOWH-AM. In the 1950s and early 1960s, AM stations had been dominant, and FM stations were usually money losers. But in the mid–1960s, that situation was on the way to being reversed: FM stations were the only ones showing growth in listenership. So while Jack Sampson was making his phone call, Bud Connell said to Robert, "We need to buy an FM station for St. Louis, or we're going to get hurt." Storz visibly stiffened, gripped the arms of his chair, and replied in the manner of a man accustomed to making snap decisions that no one questioned: "Young man, I can see that you don't know what you're talking about!" Jack Sampson returned to the table shortly thereafter, and nothing was said about the incident. Connell, however, noticed an immediate cooling and obvious inattention toward him on the part of 68-year-old Storz. When the meeting wound down and Connell left the table, Robert told Jack Sampson, "I don't like that young man. Get rid of him." Sampson resisted, and recounted Connell's high value to the company. Sampson got his way that evening, and was able to retain Connell until late 1968, when Bud Armstrong, carrying out instructions from the senior Storz, told Sampson, "It's time to fire Connell. It's either you or me."

Within a year after Todd's death, the home office had begun to take over the operations of some, but not all, of the Storz stations. Bud Armstrong, the executive vice president, was transferred from Kansas City to Omaha. Tellingly, no national program director was hired to replace Grahame Richards, whose contract had been terminated shortly after Todd's death. Bud Connell was originally being groomed for that position, but because he had very briefly made the case to Robert for acquiring FM stations, he was delisted. Even though Deane Johnson was not at that meeting, he still rankles at Robert's rebuke of Connell many decades later: "In one comment at dinner, a comment that was actually right on target, the seasoned and time-proven program director [Bud Connell] who had provided Storz Broadcasting with huge ratings success that produced massive profits for the company, was marked for elimination for no other reason than that Robert H. Storz didn't have a clue." Johnson's conclusion was that "Robert H. was exactly the opposite of Todd in operating philosophy, destroying the company in the process."

Reflecting on his tenure with the Storz organization, Bud Connell summed up his philosophy (and, he believed, Todd Storz's as well) by highlighting the difference between the narrow "business" definitions of air talent used by Todd's father, and true broadcasting innovators. Connell said,

I hired people who were "rocks," people who were strong and each one different, each one unique. We honored them, and as a company, benefitted from the distinctions among these people — and I believe Todd's philosophy was the same. Conversely, Todd's father — and the lot of "more music"— oriented, sound-alike, post–Storz broadcasters such as Bill Drake and his imitators — preferred people to be like bricks. All personnel in a given department were encouraged to be the same, patterned after the same specs, and therefore, interchangeable. Drake formatting was the inadvertent bridge from commercially-loaded Top 40 programming on AM stations, to the forthcoming low-commercial-load FM stations. Drake converted floundering major market facilities into Top 40 jukeboxes with few interruptions and "less-talk" deejays, who could be swapped into any day part among any number of stations. When pitted against each other on equal AM facilities, in the creative realm of radio broadcasting, Todd's way excelled; the new way did not — until FM reached the tipping point and dominated listening habits.

But in 1966 — two years after Todd's death — at least three of the Storz stations did implement a modified Drake format, in which the focus was on the hit record, and the disc jockey was reduced to reading short slogans. As Deane Johnson recalls,

> When Bill Stewart returned to the company for a short time, someone made a decision to copy the Drake format. Bill Stewart implemented it at WTIX, KOMA and WDGY. It was not implemented at KXOK and WHB. I was at WTIX then, and I was leaving Storz to go to Cleveland. Bill Stewart was in the process of running from market to market to implement it. They went to 20/20 news and the whole thing.

KXOK manager Jack Sampson remembers:

> Stewart came into St. Louis at about that time and he was calling the deejays — particularly Ray Otis and William A.— to come down to his hotel suite and talk to him about it. And I got wind of it — and it really ticked me off. So I called Bud Armstrong and said, "You get that — out of town and don't send him back. We don't need him, we don't want him, he's just messing everything up." So we just literally threw Bill Stewart out of town, and that was at about that time.

Deane Johnson added:

> Ironically, the same thing happened in New Orleans. Stewart came to town and was calling the jocks downtown to tell him what was going wrong at the station, and [general manager] Fred Berthelson called Armstrong and said he wanted him out of town. But then shortly thereafter, they went with the Drake format and Stewart was the guy that implemented it. I'm not suggesting that this copying of the Drake format was a bad thing — it was probably a good thing at the time. But it was a departure from the standpoint that now a programming company [Storz Broadcasting] was copying some other broadcaster.

Asked if WHB and KXOK continued with the Storz format because the

Twelve • The Legacy of Storz Broadcasting

Storz sound was more engaging and more fun, Jack Sampson replied, "We thought it had to be fun — for everybody. That was the intangible that was very important." Bud Connell added, "Drake was a music-oriented consultancy. We [Storz's KXOK] were a personality-oriented station ... a personality-oriented *corporation*."

Richard W. Fatherley remembered that for about two years following Todd's death, not much changed. But in 1966,

> they shut down the Miami headquarters and shipped it all back to Omaha. And that's when the feathers hit the fan. Memos came out: No general manager can make any capital expenditure over $5,000 without permission from the Home Office. Invoices, mail, paychecks — everything came out of the central office in Omaha.

Indeed, throughout the mid–to–late Sixties, home office edicts and reporting requirements increased exponentially. Managers lost some of their independence as Robert H. Storz installed central accounting and payables, taking it away from the managers. Life was about advertising, profits, and "the bottom line"— period. Everything revolved around money. Middle management throughout the company was purged. Programming was ignored. Jack Sampson remembers:

> As time went on, we had to submit new reports to the home office. But never on programming. They were only interested in sales. In the early '70s, they started a system where each salesperson had to submit a daily sales report to the home office, along with an estimate of how much they would bill that month. Then the manager had to submit a weekly report on sales, with a great deal of detail. I did all of it for awhile, and then quit [sending in reports] with no rebuke from the home office. But after I hired an African-American for the news department, and then hired a woman for the sales department in 1971, they really grilled me. At managers' meetings I questioned some of their ideas. Armstrong called me out and told me to quit arguing; he said I had turned into the managers' "shop steward."

In retrospect, Jack Sampson had every reason to "go to bat" for his fellow Storz station managers. For this book, Sampson produced the following comparison of the managers meetings as they were conducted while Todd Storz was alive versus how they operated when Todd's father ran them:

> MANAGERS MEETINGS AS CONDUCTED BY TODD STORZ:
> Meetings were sometimes held as often as twice a year. The atmosphere was fun and easy. We were all expected to be creative. There was no specific agenda, but Todd had a plan in mind. The emphasis was on programming. Bill Stewart was influential in the early years. Of course, selecting which records to play was essential. We used *Cashbox* and *Billboard* lists. On-air contests were a way to build audience involvement. Popular deejays gave the stations a certain "star"

quality, but managers had to limit how far they could deviate from the format. Managers meetings were held in interesting places such as Chicago, St. Louis, Miami, New Orleans, Washington, D.C., Kansas City, and New York. At the meetings, everyone presented his problems for discussion. The agenda was loose, and changeable. Occasionally, Dale Moudy would discuss new engineering innovations. There was lots of kidding and "needling." They played jokes on me several times — such as not telling me exactly where in the city the meeting would be, and I was supposed to figure it out. We played practical jokes on everyone, including Todd. Sometimes we took our wives. We had big dinners in nice restaurants. There was never a specific time to adjourn. Managers meetings usually lasted about two days. There was no drinking until after 5 P.M.

MANAGERS MEETINGS AS CONDUCTED BY ROBERT H. STORZ:

Meetings were held only once a year, always in Omaha, usually in January or February. The mood was serious, with few laughs. There was a strict written agenda, all timed to begin and end on schedule. Under Robert H., we concentrated on business. Topics could include "money-saving" (i.e., do we really need a news staff?), insurance, legal matters, dealing with the FCC, and staff assignments. There was no time for each manager to express problems. Robert H. wanted to put more pressure on sales staffs — for example, can we eliminate sales managers? He asked if a manager could carry an account list. No commissions were paid. There was no discussion of music or programming. We had to justify the position of our local program directors, and they had to be on the air, as well. The dinners were formal and were held at the Storz mansion in Omaha. Everyone dressed up during business meetings in coats and ties. All persons were addressed formally — particularly home office [Omaha] personnel. Wives were never included, except at the 25th anniversary meeting in Miami in 1968. There was no drinking, except for German wine [Blue Nun] at dinner.

It is not an exaggeration to say that Robert H. Storz never understood or acknowledged the role of programming in the success of the company when Todd was running it. Instead, he complained that the managers were spending too much on it. The senior Storz had such a programming disconnect on what his Top 40 radio stations were doing, that he once asked Jack Sampson if KXOK "could play some Lawrence Welk from time to time, or a lively Sousa march in the morning." The managers humored him, and did whatever they could to keep him away from the established programming elements that had made the Storz organization successful.

Bud Connell wasn't the only one to get on the wrong side of an intractable Home Office on the subject of owning FMs. Jack Sampson also championed the acquisition of FM stations. In 1971, a Class C (high power) FM licensed to St. Louis was available for under $150,000 (about $833,107 in 2012). A full power FM in a top ten market for less than a million dollars was rare indeed, but Jack didn't get a reply to his letter from the Storz home office, and later Bud Armstrong told him to quit "pestering" them about FM. (The St. Louis

FM station later sold for millions.) Armstrong told Sampson, "If you want an FM, why don't you buy your own!" Jack wasn't in a position to do that at the time. But it is clear that investing in FM stations was one huge opportunity that wasn't merely missed — it was summarily dismissed by Robert H. Storz.

The first "flat" year for KXOK's profit came in 1974, after ten years of 15 to 20 percent annual increases. St. Louis FMs were continuing to move up strongly in the ratings, so in Sampson's view, the fact that KXOK had managed not to *lose* money to the FMs was a positive. But starting in 1973, the home office assigned Jack Sampson to manage WTIX in New Orleans ("because of big problems there"), as well as checking in on WHB in Kansas City (because Sampson was a corporate officer for Missouri) — all in addition to running KXOK and negotiating with unions. After more than 20 years with the Storz organization (Sampson had joined the company at WHB back in 1953), he and his wife Arvy were finding the weekly travel and responsibilities "crushing." Jack remembers "They [Storz] were really unhappy with me. So that's when I decided to leave. There was no tomorrow for me." By December of 1974, Sampson was actively looking for his own FM station, and he bought his first one in Hutchinson, Kansas. Sampson left KXOK in the spring of 1975. He said, "I saw Armstrong at a WHB reunion in 1978, and he seemed somewhat bitter, but we didn't talk much."

To be fair, it wasn't only Robert H. Storz's refusal to buy FM stations that put the Storz stations into decline. New cable TV systems were invading cities and towns across America, offering not just clearer pictures and more program choices, but also "niche" channels for narrowly targeted audiences. One of those was MTV — "music television" — which played hit records accompanied by videos that often were more compelling than the music. The nationwide service signed on August 1, 1981. The very first music video shown on MTV wasn't a new record, but rather the 1979 hit by the British group The Buggles titled "Video Killed the Radio Star." The tune described a singer from the Golden Age of radio whose career ended with the advent of TV — an obvious parallel to what was happening at that very moment to AM hit music stations like those owned by Storz.

After years of declining audiences and revenues, all six of the Storz AM stations were sold between April of 1984 and September of 1985 for prices that were far lower (especially when taking inflation into account) than they would have fetched in their heyday under Todd's programming-oriented guidance. Deane Johnson speculated that when the stations were being sold off in 1984–85, Storz Broadcasting could have been a $100 million company instead of a $20 million company — had they bought the FMs which Robert H. Storz refused to consider. Instead, the sale prices were as follows:

WTIX, New Orleans— sold June 1984 to Price Communications Corp.—$3 million ($6,494,870 in 2012).

KOMA, Oklahoma City— sold June 1984 to Price Communications Corp.—$3 million ($6,494,870 in 2012).

WDGY, Minneapolis— sold October 1984 to Malrite Communications Group — just under $3 million.

WHB, Kansas City— sold April 1985 to Shamrock Broadcasting (a Walt Disney co.) — sale price not found.

KXOK, St. Louis— sold August 1985 to Chester Broadcasting Co.—$2 million ($4,329,913 in 2012).

WQAM, Miami— sold September 1985 to Sunshine Wireless Co—$2.85 million ($6,170,126 in 2012).

With all of its stations sold, the Storz Broadcasting Company ceased to exist. That had been the station group's name since 1958 — one year after Todd Storz sold his original KOWH, Omaha, for a record price and began adding other stations to his roster of holdings.

The one person who was "in on the ground floor" with Todd in the early days of KOWH, and who much later presided over the liquidation of the Storz Broadcasting Company in 1984–1985, was Bud Armstrong, who died in October of 2003 in his hometown of Omaha. His funeral, like Todd's, was attended by former Storz station managers and dozens of staff members. Armstrong's insistence on the primacy of programming that attracted and held listeners was chief among his legacies, and was one of the key elements in the success of the Storz stations prior to Todd's death.

Jack Sampson offered some sobering words on what transpired after Storz's demise:

> After Todd died, which was shortly after I got to KXOK, the Top 40 formula almost never changed after that — clear up to the end when they were getting eaten up by FMs. A lot of corporations get that way. You get a formula that works, and you like it and understand it and live it, but you've lived with it so long that other people are doing things better. If you don't change with the times ... that was a lot of Storz's problem [in the later years under Robert H.] After Todd died, they did no more upgrading or buying — especially not FM's. But the formula worked well, into the late 1960s, when the FM's began to get big.

The decline of the Storz Broadcasting Company is sad because — as the saying goes — "it didn't have to be this way." If FM facilities had been acquired, and if inventive program directors and managers had been supported, the company might have continued to be a programming innovator, and thus retain its audiences. Instead, there was a span of about 25 years (1949 to 1974–5) when Storz stations were the epitome of innovative, memorable, and delightful radio broadcasting — and then came the slow decline.

Anyone who tunes across the AM or FM radio bands now, expecting to

Twelve • The Legacy of Storz Broadcasting

find a station that sounded like the ones operated by Storz during those halcyon days would be sorely disappointed. Music has mostly disappeared from AM radio, replaced by talk and news. The FM stations that now carry music probably get it from a syndicated programming service — often delivered by satellite — and there is seldom much of an attempt to localize it. The only local elements likely to be heard are the commercials, a very limited number of station identification jingles repeated *ad nauseam*, and perhaps news headlines and weather — often recorded. There are very few local disc jockeys still practicing, and even fewer true "air personalities." Sirius/XM satellite radio carries "oldies" channels that offer occasional re-creations of Top 40 as it was heard in its heyday, but syndicated Top 40 "countdown" shows like those once hosted by Dick Clark and Casey Kasem are gone. Today, popular music is programmed for musical subgenres such as country, rock, R&B/hiphop, Latin, pop, dance/club, and more. Many stations shy away from playing any music that isn't already a well established hit, or an oldie. So it is not surprising that as of this writing in 2012, people are either listening to music online through services such as Pandora and Last FM, or are downloading songs that they might have originally heard on television, at the movies, or in a web video. There are few jingles and disc jockey personalities to be heard because this generation cannot imagine how those elements might actually add to their enjoyment of the music. As Bud Connell wrote in an article titled "The Radio Generation and the Origin of Top 40,"

> In the years after 1970 ... "more music" programming took the listener's spotlight off the station's personality and placed it on the individual record, the hit song. Listeners began to control the medium through rapid station-changing, and seeking the better record from moment to moment. After all, music was now all they had to listen for.

Connell's description above of what happened when the "more music" Drake format swept the radio airwaves and listeners had lots of station choices was very different from the formative years of the Top 40 format (from the mid- to late 1950s), when the rock-and-roll music that younger listeners wanted to hear was scarce. In those days, the one dependable place to hear it was on a Top 40 station. As Connell says,

> No one, regardless of age, escaped the influence of Top 40 Radio. The over–65s listened and put it down; the 50–64s listened and tolerated it; 25–49s listened habitually especially in the drive times, midday, and overnight; 18–24s listened in the afternoons and evenings; and the under-18 crowd were devotees whenever they were awake.

The late 1950s was also a time when television networks had absorbed just about every network radio program that could reasonably be restaged for the home screen. As a consequence, radio stations were forced to do more local programming than at any time since radio's beginnings in the 1920s. The most successful Top 40 stations were usually ones that didn't just *originate* from a certain city — they *featured* that city in their newscasts, public service announcements, commercials, and disc jockey patter. Disc jockeys hosted local teen dances and concerts. Contests were organized to award prizes to local citizens. Station ID jingles kept emphasizing the positive aspects of the metropolitan area that the station was licensed to serve. In those days, watching television took you away to somewhere else, but listening to Top 40 radio immersed you in the place where you lived.

Professional radio broadcasters in the Top 40 era were in general agreement that the Storz stations were the leaders in developing, implementing, popularizing, selling and defending the Top 40 format. The McLendon station group — whose KLIF in Dallas was the "laboratory" equivalent of Storz's KOWH in Omaha — also had a significant impact. Like Storz Broadcasting, the McLendon stations spent lavishly to promote their stations, and Gordon McLendon was a staunch promoter of the format. But he also was interested in experimenting with stations offering "beautiful music," "all-news" and even "all-ads" programming. And he produced two low-budget horror movies.

Todd Storz remained focused on creating highly entertaining hit music radio stations. At the stations owned and operated by Mid-Continent Broadcasting Company, and later the Storz Broadcasting Company, the bedrock principle was to get the programming right so that advertising sales would result. It seems obvious now, but in the early days of the Top 40 era, other radio stations were still airing individual programs rather than a consistent sound all day long — and television had been selling *only* individual programs since its beginning. Today, a "news junkie" can tune in CNN and leave it on day and night, because it provides the constant variety and updating of whatever is happening in the news — all formatted in familiar patterns, and hosted by attractive presenters. Bud Connell supports a view, which is also held by Deane Johnson — namely, that the Storz format is alive today in the formatting and presentation of cable news networks. As Johnson sees it, cable news programs "hit the popular stuff, they repeat it over and over, and it is formatted just like the format of a Top 40 station 40 years ago." Connell agrees and offers a more detailed description:

> They've got it all: the individual news story packages (analogous to Top 40's records), the hosts (like Top 40s deejays and news voices), the promos between news packages, the commercial stop-sets, the promotions often tied to certain personalities or shows such as *Fox and Friends*, and the "processed" sound which includes musical intros and outros, and even musical punctuation.

Johnson's and Connell's observations that cable news is the current incarnation of the Top 40 format seem to be on target. But it's nowhere near as much fun.

It is worth asking if the management practices and "philosophies" that were developed by the Storz stations have remained operative in today's broadcasting industry. Richard W. Fatherley produced a list of the most important operating and business priorities at the Storz stations in the mid–1960s, when their procedures were fully established and widely imitated. The Storz stations' top 5 "Operating Priorities" had been supplied to Fatherley by Bud Connell at KXOK in 1964:

1. License Security (to maintain FCC authority to operate the station).
2. Programming and Station Personalities (to attract, build, and maintain audience).
3. Audience (for sale to advertisers).
4. Sell and market the station's audience (to generate revenue).
5. Station Personnel (for day-to-day station operation).

A year later in 1965, Fatherley interviewed Steve Labunski who was then president of NBC Radio, following a stint at WMCA in New York. (Formerly, Labunski had been sales manager at WHB in Kansas City, and general manager of WDGY in Minneapolis.) He offered this list of the Storz stations' top 5 "Business Priorities":

1. Build and maintain programming to attract audience.
2. Measure the audience.
3. Convert those audience measurements into a set of economic values (gender, age groups, income groups, households, and employment categories).
4. Sell those economic values to advertisers.
5. Generate revenue to operate and promote the radio station's business priorities.

Although Connell's list began with "license security" at the top — probably because of the close scrutiny the FCC had given to Storz's station acquisitions to determine if they were in the public interest — his list and Steve Labunski's are strikingly similar. Connell's number 2 point, and Labunski's number 1 point focused on building and maintaining programming, which would in turn build and maintain audience. That was always a top priority of the Storz Broadcasting Company while Todd Storz was alive. Measuring the size and composition of the resulting audience was next, so that the audience's economic value could be made clear to advertisers. The axiom, "Get the programming right and the sales will follow," might not have been originated by the Storz stations, but as an operating philosophy, it was closely adhered to.

During its heyday, Top 40 wasn't just one format among many — as eventually became the case. Instead, it was *the* dominant radio format, and the Storz stations were its principal innovators. Bud Connell believes that if the Storz Top 40 format — as exemplified by the early WDGY or the later WHB or ultimately KXOK — were recreated today on FM stations with adequate power, good dial positions, and competitive commercial loads, those FM outlets could again dominate their markets. But that seems unlikely to happen in an era when radio broadcasting must compete for audiences with television and a galaxy of online offerings.

So, did Storz programming and management practices have a "lasting impact" on the evolution of radio broadcasting? The absence of a fully realized Top 40 format on the modern radio dial is a silent answer in the negative. The entire radio industry is challenged by the availability of music, news, and personality "utilities" all vying for attention on your computer or your smart phone. The person making instantaneous choices about web content has become the programmer of his or her own media experience. The notion of waiting for hours to find out which record is ranked number one, and the joy of hearing it at last, is unthinkable: Today, you merely choose your music genre, and just one click of a mouse or flick of a finger can show you a song's popularity, and play the tune. Given how "empowering" the experience is for the online user, there is probably no going back.

But in the heyday of the Mid-Continent Broadcasting Company, and later the Storz Broadcasting Company, Top 40 programming was the most fun you could have with your ears. For many of us who remember how good it sounded, and how it made us feel, music on radio today is at best a pale imitation of what we know the medium once offered, and probably never will again.

Appendix:
A Storz Broadcasting Timeline
By Bud Connell

Where no month abbreviation precedes an entry, that entry has been placed in the most likely chronological position.

1949
April: Storz, operating as Mid-Continent Broadcasting, purchases KOWH AM and FM in Omaha for $75,000. Todd Storz becomes vice president and general manager; Robert H. Storz, president. With the purchase, they inherit George "Bud" Armstrong, Dale Moudy and Jack Sandler.

1950
Don Loughnane and Sandy Jackson join KOWH.
March: KOWH begins 4 P.M. 90-minute *Sweet Music* current pop record show.
Local audience survey reveals *Sweet Music* show has high ratings while all other "block" programs show small audience ratings and no growth.
Current pop music expanded on KOWH.
December: Popular music now scheduled from noon until signoff.

1951
Virgil Sharpe joins KOWH.
News "live at 55" begins on KOWH.
August: First record released containing the terms "rock" and "roll" in the lyrics; "Sixty Minute Man" by The Dominos. Played mostly by R&B stations.
October: First KOWH "Treasure Hunt" conducted; ties up traffic.

1952
Top 10 introduced on KOWH; popular music now scheduled all day.
Audience promotions begin on KOWH.
KOWH's Hooper ratings exceed 36 percent share of audience.

1953

May: WTIX, New Orleans, purchased for $25,000. Bud Armstrong transferred from KOWH and named WTIX general manager.

WTIX's *Top 40* show is officially named by Bud Armstrong and is aired on a daily schedule.

1954

March: WHB, Kansas City, is purchased for $400,000. Storz inherits Jack Sampson and his staff with the station purchase. Bud Armstrong is transferred from New Orleans and named WHB general manager. Studios move to the Pickwick Hotel, with new equipment developed and installed by Dale Moudy.

July: Top 40 begins airing on WHB.

August: KOWH achieves 48 percent share of Omaha audience.

Todd Storz and Dale Moudy invent the reverberation unit for radio broadcasting.

Dale Moudy invents the nation's first telephone answering device for the KOWH "Call Santa" promotion.

1955

January: Mid-Continent Broadcasting announces Robert H. Storz becomes chairman of the board, Todd Storz becomes president, Virgil Sharpe becomes KOWH vice president and general manager, and WHB manager Bud Armstrong is named a corporate vice president.

Gordon McLendon visits Kansas City, monitors WHB, and takes the format sound back to install on his KLIF in Dallas.

December: WDGY, Minneapolis, is purchased for $334,000 or $212,000—sources disagree.

1956

January: Steve Labunski is moved from Omaha to become general manager of WDGY. Don Loughnane assigned program director of WDGY. William L. "Bill" Armstrong is transferred from WTIX to WDGY as program director. He will ultimately become a United States Senator from Colorado.

May: WQAM, Miami, is purchased for $850,000.

May: Storz stages now-famous "treasure hunts" in Omaha and Minneapolis.

June: Jack Sandler is transferred from KOWH to WQAM as general manager.

June: Todd Storz and Top 40 programming are skewered in scathing *Time* magazine article.

June: FCC says Storz contests featured in *Time* magazine raises "public interest" questions. The FCC threatens to hold a hearing on the WQAM transfer.

June: Bill Stewart joins KOWH as program director.

June: Todd Storz "discovers" Bud Connell on an Arkansas station and offers a disc jockey position at KOWH. Connell doesn't move—yet.

June: Jack Sampson is promoted to WHB sales manager.

August: KOWH Appreciation Show featuring ten pop stars is held in Omaha.

1957

Herb Oscar Anderson leaves Storz to join Jim Backus and Merv Griffin on ABC Radio in New York.

February: KOWH Staff lists Virgil Sharp as VP & GM, Grahame Richards as announcer, and Bill Stewart as program director.
March: Bud Connell joins KOWH as on-air personality, one month before the station is sold.
April: Storz sells KOWH for $822,500 – a record for a daytime-only station — pending FCC approval.
May: Todd Storz and his station group are a feature article "The Storz Bombshell" in *Television Magazine*.
June: Steve Labunski leaves Storz to become vice president of ABC Radio, and later president of NBC Radio.
June: Jack Thayer named general manager of WDGY.
June: Bill Stewart named Storz national program director.
August: Dale Moudy resigns Storz to join ABC Radio in New York.
November: KOWH ownership officially transfers to William F. Buckley, Jr.
November: Bud Connell hired by Gordon McLendon for father-in-law's WNOE New Orleans to program against Storz's WTIX.

1958
March: First Storz Disc Jockey Convention — Kansas City.
May: Mid-Continent Broadcasting Company changes name to Storz Broadcasting Company.
June: WTIX, New Orleans ratings falter from Connell programming WNOE.
Steve Labunski hires WHB's Ruth Meyer to program WMCA New York.
November: KOMA, Oklahoma City, is purchased for $600,000. Jack Sampson named vice president and general manger.

1959
Jack Thayer leaves WDGY to become a Metromedia executive, and later president of NBC Radio.
April: Todd Storz and Bud Armstrong interview Bud Connell for WDGY general manager position, but do not hire him because of his youth.
May: Second Storz Disc Jockey Convention, Miami. Bill Stewart forgets to leave passes to the Peggy Lee convention finale show for Todd Storz.
May: Bill Stewart exits Storz Broadcasting for the first time.
Grahame Richards hired and named Storz national program director.

1960
August: KXOK, St. Louis, is purchased for $1.5 million.
December: Bud Connell exits WNOE, New Orleans to manage and program WFUN, Miami against Storz's WQAM.

1961
March: Todd Storz announces move of corporate offices to 767 41st Street, Miami Beach, Florida.
May: KOMA installs Shafer automation system.
June: Bud Connell is hired to program KXOK after decimating WQAM's ratings. New personnel hired, 12 months of promotions recorded and scheduled.
July: Danny Dark is imported to KXOK from WFUN, Miami.

December: KXOK achieves number one ratings in Nielsen, going from 4 percent to 20 percent share in first sixty days after re-staffing and re-programming.

1962
July: KXOK achieves number one ratings in Pulse, a "weighted" survey.

1963
Danny Dark departs KXOK. Becomes signature voice for NBC-TV for almost thirty years.

1964
April: Jack Sampson transferred from KOMA to KXOK as general manager, replacing C. L. "Chet" Thomas.

April 13: Todd Storz's sudden and untimely death; Bill Stewart returns to Storz as WHB Promotion and Publicity Manager on the same day as Todd's death.

April: Bud Connell hires Richard Ward Fatherley as KXOK production director.

1965
June: KXOK helps promote Frank Sinatra's *Rat Pack Special* benefit show

1966
July: KXOK achieves status as the number one independent station status in entire USA according to The Pulse, Inc.

August: KXOK sponsors The Beatles' appearance at St. Louis's Busch Stadium.

October: Bill Stewart exits Storz Broadcasting for the second time.

1967
Richard "Dick" Harris exits Storz to become president of Westinghouse Broadcasting.

June: Richard Ward Fatherley is transferred to WHB to be program director.

Bud Connell asks Robert H. Storz to acquire FM facilities and is rebuffed.

1968
November: Bud Connell is fired from KXOK upon orders from Robert H. Storz.

1970
October: Bud Connell consults/programs WFUN, Miami, against Storz for the second time and compromises WQAM's audience ratings yet again.

1971
Jack Sampson asks Robert H. Storz to acquire an FM station for St. Louis. Through Bud Armstrong, the request is rebuked by the elder Storz.

Jack Sampson is assigned oversight of WTIX in addition to KXOK.

Jack Sampson exits Storz and purchases his own FM station.

1982
May: WABC New York ends more than twenty years as a Top 40 station.

1984
June: WTIX sold to Price Communications for approximately $3 million.

June: KOMA sold to Price Communications for approximately $3 million.
October: WDGY sold to Malrite Communications for approximately $3.5 million.

1985
April: WHB sold to Shamrock (Walt Disney) for approximately $3.5 million.
April: KXOK sold to Chester Broadcasting for $2 million.
June: WQAM sold to Sunshine Wireless for $2.85 million.
Gross sale price of all six Storz stations was $17.6 million, according to Herb Engdahl of the Robert H. Storz Foundation.

1989
February: Todd Storz inducted posthumously into the Nebraska Broadcasters Association Hall of Fame.

1996
July: George "Bud" Armstrong interviewed by University of Maryland for the Library of American Broadcasting.

2001
June: Bud Connell inducted into the St. Louis Radio Hall of Fame with Harry Caray and Paul Harvey, both pre-Storz KXOK employees.

2004
June: "Top 40 is 50" commemoration is held in Kansas City with retired Senator and former WTIX Program Director Bill Armstrong as keynote speaker.

2006
June: Richard Ward Fatherley is inducted into the St. Louis Radio Hall of Fame.

2008
June: Jack Sampson is inducted into the St. Louis Radio Hall of Fame.

Chapter Notes

Chapter Two

1. A "kolach" is a type of pastry, popular among Omaha's Czech and Polish communities.

2. The term "disc jockey" had been coined by *Variety* magazine in 1941; it was applied to people who played records to an audience — both at social gatherings and on the radio.

3. Listen to the brief clip of Sandy Jackson on KOWH at *www.deanejohnson.net/audio/KOWH Sandy Jackson.shtml*. In addition, a one-hour recording of Sandy Jackson made after he left KOWH can be heard (by subscription) on *www.reelradio.com*. Search for "Sandy Jackson, KOIL, Omaha NE March 7, 1962."

4. Such enormous popularity would later raise the question of why Jackson wasn't moved to another Storz station when KOWH was sold in 1957. One answer was that he was too valuable to KOWH and thus might disrupt the sale. Another was that Jackson was "a local phenom," a term used by program director Bill Stewart to describe a disc jockey who gained a great following in a given market, but who would not likely succeed elsewhere because his local associations were so important. Yet a third theory was that Jackson, who had a large family and deep roots in Omaha, just didn't want to move.

Chapter Three

1. Go to *http://www.walkerpub.com/radio_thrutheyears.html* to read Walker's full essay, "Thru the Years...."

Chapter Four

1. In 1998, the station swapped frequencies with KCMO-AM on 810 kHz to achieve a larger daytime coverage area for its sports-talk format.

2. The intersection of "12th Street and Vine" was celebrated as the nexus of the Kansas City jazz scene in the song "Kansas City," which had been written by Jerry Leiber and Mike Stoller in 1951, and became a huge Top 40 hit for Wilbert Harrison in 1959.

3. This and some other details of Jack Sampson's years at WHB are from a telephone interview recorded by Prof. Tom McCourt on June 3, 2003, which was acquired by Dick Fatherley for possible use in this book.

4. One of the elements that made WHB (and all Storz stations) "sound darn good" was their extensive collection of station identification jingles. Follow this link to hear 14 minutes (!) of WHB jingles from the 1950s and 1960s, and to appreciate how much music there was between the records. Go to *http://soundcloud.com/randb/whb-jingles-compressed-file*.

Chapter Five

1. A 1970 University of Minnesota master's thesis, "The History of Radio Station WDGY," by Jerry Verne Haines was the source for key details on the early (pre-Storz-ownership) history of WDGY. As of 2012, the entire thesis is available to read — and WDGY airchecks can be listened to on-line — at *www.radiotapes.com/WDGY.html*.

2. You can hear Deane Johnson's last air shift on KDWB — just before he became the station's full-time general manager. Go to *www.deanejohnson.net/audio/KDWB_Deane _Johnson_Air_Check5-21-70.*

3. Dave MacFarland acknowledges that it was Charlie Murdock's afternoon show on WQAM in the early 1960s that got him to think about a career in radio. "In the early 1960s, WQAM was by far the most exciting station on the radio, and Charlie Murdock's afternoon show had all the energy and fun that you wished could somehow be in your own life." Richard W. Fatherley arranged for MacFarland to meet Charlie Murdock just a few years before Murdock's death. A brief aircheck of Murdock on WQAM in April 1962 is available at *www.560.com.*

4. You can listen to a recording of "WFUN Fundamental News" from October 23, 1961, supplied by Deane Johnson. Go to *www. deanejohnson.net/audio* and scroll down the list of Audio Archives to the heading "WFUN–Miami."

Chapter Eight

1. The former Mid-Continent Broadcasting Company changed its name to Storz Broadcasting company in May 1958.

2. Williams's comment was broadcast over KOMA-FM during its April 17, 2003, "Fan Jam" weekend commemorating the Great Plains' biggest Top 40 radio war — between KOMA and WKY.

3. The MacKenzie Repeaters and other specialized audio production devices are explained in greater detail in Chapter Ten on "Elements of the Storz Station 'Sound.'"

4. There is more detail about radio program automation systems and their effect on Storz programming in the chapter on "Elements of the Storz Station 'Sound.'"

Chapter Nine

1. As of mid–2012, an excellent website for all things KXOK can be found at *www.*

630kxok.stlmedia.net. There are several KXOK airchecks from the 1960s, plus information on other Storz stations and Top 40 in general. Especially recommended is "KXOK was St. Louis' 'American Graffiti'" by St. Louis media historian Frank Absher.

2. The August 29, 1960, edition of *Billboard* magazine said the sale price was $1,500,000 (about $11,398,936 in 2012), but that total included "a square city block in downtown St. Louis and 90 acres at the transmitter site."

3. A selection of KXOK jingles is at *http: //630kxok.stlmedia.net/audio/jingles/index.htm.*

Chapter Ten

1. PAMS jingles can be listened-to in decade-by-decade "samplers" or by individual jingle "cut" [track] at *www.pams.com/ listen.html.*

2. See "The WABC PAMS Story" for more detail about the PAMS jingles that were also used at all of the Storz stations. Go to *www.musicradio77.com/jingles.html* and scroll down to the heading "The WABC PAMS Story." Click on that heading.

3. A radio station's cumulative audience (also known as "reach") is the number of *different* people who listen to the station for a period longer than the basic time unit — usually 15 minutes.

4. An excellent feature on the MacKenzie Repeater — forerunner of the broadcast cartridge tape recorder/player — is at *www.reel radio.com/reports/mackenzi.html* (Note there is no final "e" in the mackenzie name in the URL.) Viewing this report does not require a subscription.

5. From "WQAM's 'Big Brother' Is Cut Down to Size," an article which was reproduced from the *Miami Herald* and posted on *www.560.com.*

A Bibliographic Note

This book is comprised primarily of original material, supplied by the authors and by Bud Connell and Jack Sampson, whose firsthand experience in Storz station broadcasting was extensive and sharply recalled. Authors Fatherley and MacFarland have published previous books on radio history:

Fatherley, Richard W. *Radio's Revolution and the World's Happiest Broadcasters,* 1990s.

In the late 1990s, Dick Fatherley wrote and self-published a short paperback history of the Storz stations titled *Radio's Revolution and the World's Happiest Broadcasters*. He made it available to anyone for the cost of postage. He also narrated a shorter version of his book, which appeared initially on audio cassette and later on CD. He also distributed those at no cost. The short printed history became the basis for the earlier chapters of the present book, and also supplied details for the chapters in this volume on WHB and KXOK, where Fatherley worked as a production director and on-air talent.

MacFarland, David T. *The Development of the Top 40 Radio Format.* New York: Arno Press, 1979. Based on Ph.D. dissertation, University of Wisconsin, 1972.

Brief quotations have been taken from: Fisher, Marc. *Something in the Air.* Random House, 2007.

Index

Numbers in **_bold italics_** indicate pages with photographs.

ABC Radio Network 29, 117
ABN (American Broadcasting Network) 89–90
Absher, Frank 125
acetate (lacquer-coated aluminum) recording discs 21
Advertising Age 79
AFTRA (American Federation of Television Radio Artists) 150
airchecks 136, 137
Alan Courtney talk program 111–112
all-hit-records music format 41
American Bandstand TV show 44, 108
Anderson, Herb Oscar 72, 138, 171
AQH enhancers 146
Arbitron ratings 75
Armstrong, George W. "Bud" 20, 22, 38, 41, 42, 51, **_58_**, 62, 82, 94, 122, 126, 144, 171–177, 183, 188
Armstrong, William L. "Bill" 38, 70
Army Signal Corps 15
Associated Press news printer 120
"At Your Service" programming (KMOX) 124
audience surveys 5
autopsy (Todd Storz) 178–179, **_180_**
Avery, Gaylord 23
Ayres, Edward B. 12, 14

Bauman, Joseph S. 9
Bay of Pigs (Cuba) 86–88
Beatles 134
Bennigan, Germany 8
Benny, Jack 138
Berthelson, Fred 38, 51, 53, 82, 89
Bill Haley and the Comets 63–64
Billboard record rankings 44
Block, Martin 23, 93

Blore, Chuck 29
Bob & Ray 138
"Booze, Broads, and Bribes" (1959 Disc Jockey Convention headline) 106
British Broadcasting Corporation productions 23
"Bruno" (Johnny Rabbit's alter-ego) 145
Buckley, William F. 47
Burkhart, Kent (WQAM program director) 81

car radios 17
Carson, Johnny 16, 27
The Cash Box record rankings 44
Central Intelligence Agency 86
Cervantes, Alphonse (St. Louis mayor) 134
charity 166
"Charlie Cherokee" 50
"Chickenman" 145
Choate School (Choate Rosemary Hall) 12
Clark, Dick 108, 148, 189
Class IV AM station 37
"clear channel" station 20
coaxial cables (TV signals) 30
"cold war" 29, 114
Columbia Brewery 9
commercial announcement limits 36
commercial production music 145
Connell, Bud vii, xvi, 6, 45–51, 85–86, 92, **_127_**, 138, 143, 147, 180, 191
contests and promotions 78–79, 146
Cook Paint and Varnish Company 57
Cooley, Lou 134
Count Basie's studio band (WHB) 56
CRC jingles 144
Creston, Iowa 1
Crowley, Mort 145

201

Index

cryptographer school 14
cume boosters 146

Dan, Chuck 120
Dark, Johnny 128
daytime-only stations 19–20
"dead air" 153
death of Todd Storz (April 13, 1964) 134, 178–182
Development of the Top 40 Radio Format xvi, 6
Devon, Delcia 135
"Dinner with Drac" 72
discs, acetate 149
divorce (Elizabeth Ann Storz) 126
"Doctor of Top 40" 3
"doctored" airchecks 45
"a dollar a holler" (inexpensive commercial) 38
Drake format 75, 184–185, 189
Drifmeyer, Leanna 23; see also "Kitchen Klatter"
"drop-ins" 145
Dunning, John 40
duopoly policy 76
"The Dynamic Change in Radio" 17

"eargasm" 153
effects: microphone filter, reverb, echo 141–142
Ekberg, Roy 14–15
Elz, Ron 133
"English (British) Invasion" 134
"Every home is a studio, and every telephone is a microphone" (Nite Beat program slogan) 113
exotic/parody/lampoon commercials 128

"failed money" 135
Fatherley, Richard W. (Dick) v, vi, xv, 3, 5, 22, 41, 63, 132, *133*, 135, 148, 151, 168, 178
F.C.C. 1, 7, 76–80
Ferguson, Art ("Charlie Tuna") 120
Fidelipac tape cartridge system 150–151
First Class License 1, 113, 118
"The First Five," KWK, St. Louis 41
Fisher, Marc (author, *Something in the Air*) 45, 60–61, 109, 128
flood re-broadcast 168
FM radio growth 18
Ford 95
formula radio xv, 6, 30
forty-five r.p.m. single records 44
Freberg, Stan 138
free money 7

Freed, Alan 108
"fun" radio 152

Hall, Claude 29
"Ham" radio operator 11
Hammond organ reverberation unit 119
Harris, Dick (WDGY general manager) 73–74
Hi-Lo contest 121
Hirt, Al 53
hit records 41, 43
Hurricane Betsy 54

IBEW (International Brotherhood of Electrical Workers) 150
illegal aliens 8
Irwin, Jim 134
It's a Mad, Mad, Mad, Mad World 7

Jackson, Sandy 27, *28*, 29, 58, 138, 158
Jacobson, Tom (Nite Beat host) 112
jingles 142–145
Johnson, Deane xvi, 3, 6, 27–28, 39, 51–54, 66, 106, 114, *115*, 138, 143, 147, 183

Kasem, Casey 189
Katz, George 116
Kay (Johnny Rabbit's companion) 133
KBEQ-FM, Kansas City 69
KBON, Omaha 15, 27
KDKA, Pittsburgh 10
KDWB, Minneapolis 2, 72–73
Keyes, Don 48, 102–104
KFAB, Omaha NE 15, 23
KFNF, Shenandoah, IA 157
KFRU, Columbia, MO 125
KILT, Houston, TX 48
King Midas and the Mufflers 118
"King of Giveaway" 76–77
"Kitchen Klatter" 23, 25, 157, 158
KKJO, St. Joseph, MO 2
KLIF, Dallas, TX 48, 190
KMOX, St. Louis, MO 124, 132, 135
Knight, James L. (owner of WQAM and *The Miami Herald* newspaper) 77
KOAD-FM, Omaha 19–20, 39, 160
KOIL, Omaha 65, 74, 107
KOMA, Oklahoma City 2, 114–123, 125, 131
KORN, Fremont, NE 173
KOWH, Omaha 1, 5, 6, 16, 19–36, 63, 65, 90, 124, 127, 149, 155–162
KSIB, Creston, IA 1
KTSA, San Antonio 48
KWBW, Hutchinson, KS 15
KWK, St. Louis 41, 135, 138

Index

KXOK, St. Louis vi, 6, 51, 124–135, 137, 138
"The KXOK Millionaire" (Dick Fatherley) 132–133

Labunski, Steve 45, 61, 70, 72, 82, 112, 125, 168–170, 191
"ladder" show 129
Leeds, Mel 96–97
LeMay, Gen. Curtis E. 114
"Light and Lively" format on ABC network 170–171
limited playlist 129
limiting amplifier 20, 149
Limpander audio loudness LIMiter and exPANDER 114
Lindberg, Charles 129
"Live at :55" news on KOWH 29
Lloyds of London (prize insurer) 159
localization 30
Loughnane, Don 21, 34, 38, 112
"Lucky Hearse Number" 60
"Lucky House Number" contest 49, 60, 159, 166
"Lucky License Number" 121
Lujack, Larry 153
Lyons, Bobby 81

MacFarland, David T. xv, xvi, 3, 6
MacKenzie program repeater 119, 143, 149–150
MacKrell, Jim *see* McKay, Jim
"Make Believe Ballroom" 23, 94
"Make It or Break It" 134
Martin, Peter (KXOKs poet laureate) 134
Martin, Ralph (Nite Beat host) 112
Martin, Ron 52, 58
Martinez, Al (nighttime voice on WQAM) 87
mass audience events 147
McCourt, Tom 25, 61–62
McGrath, Peggy 22, 25, 27, 29
McGregor, Don 51
McKay, Jim 49
McLendon, Gordon 48, 69, 92, 93, 116, 190
Meeks, Bill 142–144
Mehl, Don 15
Miami, FL 5, 7, 185
Miami Herald 76
microwave relay (TV signals) 30
Mid-Continent Broadcasting Company vi, 5, 19, 172, 190
"Mile-High Weather Eye" (WNOE) 49
Miller, Mitch 94–95
Miller, Rex (KOMA general manager) 122

Minneapolis, MN 5, 7
Missouri River flooding rebroadcast 34
Mon-Key (news sound-effects machine) 119
Moudy, Dale 11, 21, 22, 24, 25, 38, 59, 82, 89, *91*, 110, 148, 149, 171
"Multi-Phone" (audio distribution amplifier) 110
Murdock, Charlie 82, **87**, 88, 90
"music-and-news" stations 18
"My Daddy Is President" 130
Mystery Voice 166

NBC radio network 15, 40, 116–117
Nebraska Broadcasters Association 155
network television 30–31
New Orleans 5, 7
New Orleans Times–Picayune 37
"News Live at :55" 29, 38
Niehaus, H. Dayton 14
Nielsen, A.C. (ratings company) 32
"Night Club of the Air" 110
"Nite Beat" 110–113
Noe, James A. 48, 92
Norwood, Helen Stacy 33–34
nuclear Armageddon 114

O'Connor Productions 145
Offutt Air Force Base 29
Oklahoma City 5, 7
Omaha Sun 19
Omaha World Herald 15, 19, 29, 32, 34, 126
O'Neill, James 23, 32–33, 158, 160
Oreck, David 53
Orkin, Dick 145
Orleans Parish School Board 39
O'Shea, Shad 49
Otis, Ray 129, 134

PAMS jingles 107, 119, 142–144
Pandora (online music service) 189
Parker, "Colonel" Tom 8
"pay for play" 108
payola disclaimer 108
"peanut whistles" 37, 39
Pearson, Johnny 27, 38, 58, 60, 65, 116
Petersen, Ruthie (Todd Storz's secretary) 7
Peterson, Val (Nebraska governor) 30
Pickwick Hotel 58
Pietromonaco, Don ("Johnny Rabbit") 132–133
Pool, Bob 138
"Pop Music Disc Jockey Convention and Radio Programming Seminars" (Kansas City) 90–95

Presley, Elvis 8
program automation 52, 122, 151–152
promotions 32–35
pulse ratings 61, 62, 130

Rabbit, Johnny (KXOK) 132–133, 145
Radio Advertising Bureau 130
radio revenues 16–18
"Radio's Revolution" vi
Ramsburg, Jim (Nite Beat host) 112
ratings books 61
Ray, Charles (newscaster) 52
RCA Victor 8
"Rear Window Sticker" contest 49
Richards, Grahame 48–50, 74, 115, 144, 148, 183
Roberts, Elzie 124–125
Roddy, Rod 119–120
Rothenbuhler, Eric W. 25
Rounsaville, Robert 84, 86, 128

St. Louis Journalism Review 125
St. Louis Star Times 125
Sampson, Jack xv, 3, 6, 56, 115, 125, 130, *131*, 134, 188
Sandler, Jack 22, 23, 81–82, 86, 151, 160
Sandy & Green jingles 143
"Sardining" Contest 49
Sarnoff, Gen. David vi
Scarritt building 56
Schafer automation system 151
Sharpe, Virgil 26, 35, 48, 82, 154–162
Shaw, Rick 81
Sherwood, Rob 75
Sigelman, Mike 73–74
"signature" jingles 143
Simons, David ("Studio Stories") 44
Sirius/XM 184
Smith, Helen Lorraine 126
Snow, Hank 8
Soderlund, Harold 15, 23, 24, 172
Something in the Air 45, 61, 109, 128
"The Spirit of St. Louis" 129, 143
station-owned characters (KXOK) 132
station "personality" 128
Sterling, Christopher H. viii
Stewart, Bill 48, 54, 79, 81, 82, 91, *92*, 93, 95–101, 107, 129, 148, 184, 185
Stitt, Wayne 1
Storz, Arthur C. 10–14
Storz, Elizabeth Ann (Trailer) (Todd's first wife) 15–26, 126
Storz, Gottlieb 8, *9*, 10
Storz, Helen Larraine (Smith) (Todd's second wife) 126
Storz, Mildred Todd 10

Storz, Robert H. 5, 9, 10, *11*, 20, 13, 131, 135, 181–187
Storz, Robert T. (Todd), death of 134, 178–182
Storz, Susan 10–12
Storz Brewing Company 9
Strategic Air Command (SAC) 10, 29, 114
Sun Records 9
"Sweet Music" program 26
syndicated "drop-ins" 145

"Tall Paul Bunyan" (WDGY teen idol) 74
technical production facilities 148–152
Television magazine 41
Thayer, Jack 72, 138
This Business of Radio Programming 29
Thomas, Chet 125, 126, 130
Time magazine 76
time tone (audible chime) 22, 148
"Top 50 plus" 138
"Top 6 Plus 30" (KXOK's 630 dial position) 129, 139
"Top Ten Hits" 29, 38, 160–161
trade magazine advertising 66–69
Trailer, Elizabeth Ann *see* Storz, Elizabeth Ann (Trailer)
Trammel, Phil, WDGY 74
tranquilizer drugs 114
transistor (portable) radios 44
treasure hunts 60, 115, 117, 159, 167–168
Tripp, Peter 58
TSL ("Time Spent Listening") 152
Tuinal 179, 180
Tuna, Charlie (Art Ferguson) 120
Tweety Bird Christmas promotion 33–34

"Uncle Phineas" 24
University of Nebraska 11, 20, 26

van Kuijk, Adreas Cornelis ("Colonel" Tom Parker) 8
Variety 44
Vogel, Lee 111–112
"Vox Jox" 29

WABC, New York 143
Walker, Bob 43
WAME, "Whammy in Miami!" 84
WCCO, Minneapolis 71–72
WDGY, Minneapolis 2, 11, 30, 70–76
WDSU, New Orleans 42, 171
WEAF, New York 20
Wehba, Dale (WJY disc jockey) 122
Welker, Sterling 23
Western Union time signal 22, 148
WFAN, New York 20

Index

WFUN, Miami 6, 49, 85, 128
WHB, Kansas City vi, 1, 2, 5, 30, 55–69, 116, 120, 125, 135, 172
WIL, St. Louis 128, 134–135, 138
Williams, Danny 117–120
"Win a Trip to the Moon or $500" (KOMA promotion) 120
WINS, New York 125
WIP, Philadelphia 115, 125
WKY, Oklahoma City 116–122
WKYC, Cleveland 2
WLS, Chicago 171
WLW, Cincinnati 88
WMCA, New York 170
WNBC, New York 20

WNEW, New York 23
W9DYG 11, 13
WNOE, New Orleans 48–51, 65, 127
WOW, Omaha 16
WQAM, Miami 6, 76–88, 126–128
WRCA, New York 20
WSMB, New Orleans 52
WTIX, New Orleans 2, 6, 7, 30, 37–54, 63, 65, 90, 127, 171–172
WWEZ, New Orleans 39, 90
WWL, New Orleans 171

Young, Adam (advertising agency) 17
"Your Hit Parade" 27, 40–43
youth market 45

www.ingramcontent.com/pod-product-compliance
Ingram Content Group UK Ltd.
Pitfield, Milton Keynes, MK11 3LW, UK
UKHW042003140426
5217IPUK00015B/950